Studies in Systems, Decision and Control

Volume 162

Series editor

Janusz Kacprzyk, Polish Academy of Sciences, Warsaw, Poland
e-mail: kacprzyk@ibspan.waw.pl

The series "Studies in Systems, Decision and Control" (SSDC) covers both new developments and advances, as well as the state of the art, in the various areas of broadly perceived systems, decision making and control- quickly, up to date and with a high quality. The intent is to cover the theory, applications, and perspectives on the state of the art and future developments relevant to systems, decision making, control, complex processes and related areas, as embedded in the fields of engineering, computer science, physics, economics, social and life sciences, as well as the paradigms and methodologies behind them. The series contains monographs, textbooks, lecture notes and edited volumes in systems, decision making and control spanning the areas of Cyber-Physical Systems, Autonomous Systems, Sensor Networks, Control Systems, Energy Systems, Automotive Systems, Biological Systems, Vehicular Networking and Connected Vehicles, Aerospace Systems, Automation, Manufacturing, Smart Grids, Nonlinear Systems, Power Systems, Robotics, Social Systems, Economic Systems and other. Of particular value to both the contributors and the readership are the short publication timeframe and the world-wide distribution and exposure which enable both a wide and rapid dissemination of research output.

More information about this series at http://www.springer.com/series/13304

Jerzy Klamka

Controllability and Minimum Energy Control

Springer

Jerzy Klamka
Institute of Control Engineering
Silesian University of Technology
Gliwice
Poland

ISSN 2198-4182 ISSN 2198-4190 (electronic)
Studies in Systems, Decision and Control
ISBN 978-3-319-92539-4 ISBN 978-3-319-92540-0 (eBook)
https://doi.org/10.1007/978-3-319-92540-0

Library of Congress Control Number: 2018942191

© Springer International Publishing AG, part of Springer Nature 2019
This work is subject to copyright. All rights are reserved by the Publisher, whether the whole or part of the material is concerned, specifically the rights of translation, reprinting, reuse of illustrations, recitation, broadcasting, reproduction on microfilms or in any other physical way, and transmission or information storage and retrieval, electronic adaptation, computer software, or by similar or dissimilar methodology now known or hereafter developed.
The use of general descriptive names, registered names, trademarks, service marks, etc. in this publication does not imply, even in the absence of a specific statement, that such names are exempt from the relevant protective laws and regulations and therefore free for general use.
The publisher, the authors and the editors are safe to assume that the advice and information in this book are believed to be true and accurate at the date of publication. Neither the publisher nor the authors or the editors give a warranty, express or implied, with respect to the material contained herein or for any errors or omissions that may have been made. The publisher remains neutral with regard to jurisdictional claims in published maps and institutional affiliations.

Printed on acid-free paper

This Springer imprint is published by the registered company Springer International Publishing AG part of Springer Nature
The registered company address is: Gewerbestrasse 11, 6330 Cham, Switzerland

Foreword

The main objective of this monograph is to review the major progress that has been made on the controllability and minimum energy control problem of dynamical systems over the past 50 years. Controllability, first introduced by Rudolf E. Kalman in 1960, is one of the fundamental concepts in the mathematical control theory. This is a qualitative property of dynamical control systems and is of particular importance in control theory. A systematic study of controllability started at the beginning of 1960s when the theory of controllability based on the description in the form of a state space for both time-invariant and time-varying linear control systems was developed.

Roughly speaking, controllability generally means that it is possible to control a dynamical control system from an arbitrary initial state to an arbitrary final state using a set of admissible controls. In the literature, there are many different definitions of controllability which strongly depend, on the one hand, on a class of dynamical control systems and, on the other hand, on the form of admissible controls.

Controllability problems for different types of dynamical systems require the application of numerous mathematical concepts and methods taken directly from differential geometry, functional analysis, topology, matrix analysis, theory of ordinary and partial differential equations, and theory of difference equations. In this monograph, the author mainly uses state-space models of dynamical systems which provide a robust and universal method for studying controllability of various classes of dynamical systems.

Controllability plays an essential role in the development of modern mathematical control theory. There are various important relationships between controllability, stability, and stabilizability of linear both finite-dimensional and infinite-dimensional control systems. Controllability is also strongly related with the theory of realization and the so-called minimal realization and canonical forms for linear time-invariant control systems such as the Kalman canonical form, Jordan canonical form, or Luenberger canonical form. Moreover, it should be mentioned that for many dynamical systems there exists a formal duality between the concepts of controllability and observability.

It should be pointed out that controllability is strongly connected with the minimum energy control problem for many classes of linear finite-dimensional, infinite-dimensional dynamical systems, and delayed systems, both deterministic and stochastic. The minimum energy control problem for the first time has been formulated and solved by Prof. Jerzy Klamka in his survey papers.

Finally, as it is well known, the concept of controllability has many important applications not only in control theory and systems theory but also in such areas as industrial and chemical process control, reactor control, control of electric bulk power systems, aerospace engineering, and recently in quantum systems theory.

The minimum energy control problem is strongly related to the controllability problem. For controllable dynamical system, there generally exist many different admissible controls which control the system from a given initial state to a final desired state at a given final time. Therefore, we may look for an admissible control which is optimal in the sense of the quadratic performance index representing energy of control. The admissible control, which minimizes the performance index, is called the minimum energy control. This monograph contains solutions for the minimum energy control problems for many linear dynamical systems, mainly with delays in control and stochastic. The monograph contains some essential original results of the author, most of which have already been published in international journals. Due to its broad coverage and originality. It will certainly be welcome by the control community.

Gliwice, Poland
May 2018

Prof. Tadeusz Kaczorek
Fellow of IEEE
Silesian University of Technology

Preface

The main objective of this article is to review the major progress that has been made on controllability of dynamical systems over the past number of years. Controllability is one of the fundamental concepts in mathematical control theory. This is a qualitative property of dynamical control systems and is of particular importance in control theory. Systematic study of controllability was started at the beginning of 60s in the last century, when the theory of controllability based on the description in the form of state space for both time-invariant and time-varying linear control systems was worked out.

Roughly speaking, controllability generally means that it is possible to steer dynamical control system from an arbitrary initial state to an arbitrary final state using the set of admissible controls. It should be mentioned that in the literature there are many different definitions of controllability, which strongly depend on one hand on a class of dynamical control systems and on the other hand on the form of admissible controls.

Controllability problems for different types of dynamical systems require the application of numerous mathematical concepts and methods taken directly from differential geometry, functional analysis, topology, matrix analysis and theory of ordinary and partial differential equations, and theory of difference equations. In the paper, we will use mainly state-space models of dynamical systems, which provide a robust and universal method for studying controllability of various classes of systems.

Controllability plays an essential role in the development of modern mathematical control theory. There are various important relationships between controllability, stability, and stabilizability of linear both finite-dimensional and infinite-dimensional control systems. Controllability is also strongly related with the theory of realization and so-called minimal realization and canonical forms for linear time-invariant control systems such as Kalman canonical form, Jordan canonical form, or Luenberger canonical form. It should be mentioned that for many dynamical systems there exists a formal duality between the concepts of controllability and observability.

Moreover, controllability is strongly connected with minimum energy control problem for many classes of linear finite-dimensional, infinite-dimensional dynamical systems, and delayed systems both deterministic and stochastic.

Finally, it is well known that controllability concept has many important applications not only in control theory and systems theory but also in such areas as industrial and chemical process control, reactor control, control of electric bulk power systems, aerospace engineering, and recently in quantum systems theory.

Minimum energy control problem is strongly related to controllability problem. For controllable dynamical system, there exists generally many different admissible controls, which steer the system from a given initial state to the final desired state at given final time. Therefore, we may look for the admissible control, which is optimal in the sense of the quadratic performance index representing energy of control. The admissible control, which minimizes the performance index, is called the minimum energy control. Monograph contains solutions for minimum energy control problems for many linear dynamical systems, mainly with delays in control and stochastic.

The monograph contains some original results of the author, most of which have already been published in international journals.

Gliwice, Poland
May 2018

Jerzy Klamka

Contents

1	**Introduction**	1
	1.1 Controllability Concept	1
	1.2 Controllability Significance	2
	1.3 Delayed Systems	3
	1.4 Fractional Systems	4
	1.5 Positive Systems	5
	1.6 Stochastic Systems	5
	1.7 Nonlinear and Semilinear Dynamical Systems	6
	1.8 Infinite-Dimensional Systems	6
	1.9 Nonlinear Neutral Impulsive Integrodifferential Evolution Systems in Banach Spaces	7
	1.10 Second Order Impulsive Functional Integrodifferential Systems in Banach Spaces	9
	1.11 Switched Systems	10
	1.12 Quantum Dynamical Systems	11
2	**Controllability and Minimum Energy Control of Linear Finite Dimensional Systems**	13
	2.1 Introduction	13
	2.2 Mathematical Model	14
	2.3 Controllability Conditions	15
	2.4 Stabilizability	16
	2.5 Output Controllability	18
	2.6 Controllability With Constrained Controls	18
	2.7 Controllability After Introducing Sampling	19
	2.8 Perturbations of Controllable Dynamical Systems	20
	2.9 Minimum Energy Control	21
	2.10 Linear Time-Varying Systems	23

3	**Controllability of Linear Systems with Delays in State and Control**.	27
	3.1 Introduction	27
	3.2 System Description	27
	3.3 Controllability Conditions	31
	3.4 Minimum Energy Control	32
4	**Controllability of Higher Order Linear Systems with Multiple Delays in Control**	37
	4.1 Introduction	37
	4.2 System Description	37
	4.3 Controllability Conditions	40
5	**Constrained Controllability of Semilinear Systems with Multiple Constant Delays in Control**.	45
	5.1 Introduction	45
	5.2 System Description	46
	5.3 Preliminaries.	47
	5.4 Controllability Conditions	49
	5.5 Constrained Controllability Conditions for Linear Systems	50
	5.6 Positive Controllability of Positive Dynamical Systems.	55
6	**Constrained Controllability of Second Order Dynamical Systems with Delay**	67
	6.1 Introduction	67
	6.2 System Description	68
	6.3 Controllability Conditions	71
	6.4 Constrained Controllability of Second Order Infinite Dimensional Systems	77
7	**Controllability and Minimum Energy Control of Fractional Discrete-Time Systems**	89
	7.1 Introduction	89
	7.2 Fractional Discrete Systems.	90
	7.3 Controllability Conditions	91
	7.4 Minimum Energy Control	94
8	**Controllability and Minimum Energy Control of Fractional Discrete-Time Systems with Delay**.	99
	8.1 Introduction	99
	8.2 Fractional Delayed Systems.	99
	8.3 Controllability Conditions	100
	8.4 Minimum Energy Control	101

9 Controllability of Fractional Discrete-Time Semilinear Systems ... 107
  9.1 Introduction ... 107
  9.2 Fractional Semilinear Systems ... 107
  9.3 Controllability Conditions ... 109

10 Controllability of Fractional Discrete-Time Semilinear Systems with Multiple Delays in Control ... 117
  10.1 Introduction ... 117
  10.2 Fractional Semilinear Systems with Multiple Delays in Control ... 117
  10.3 Controllability Conditions ... 119
  10.4 Controllability of Discrete-Time Semilinear Systems with Delay in Control ... 121
  10.5 Controllability Conditions ... 123

11 Stochastic Controllability and Minimum Energy Control of Systems with Multiple Variable Delays in Control ... 129
  11.1 Introduction ... 129
  11.2 System Description ... 130
  11.3 Stochastic Relative Controllability ... 138
  11.4 Stationary Systems with Multiple Constant Delays ... 140
  11.5 Systems with Single Time Variable Delay ... 147
  11.6 Minimum Energy Control ... 152
  11.7 Minimum Energy Control of Stationary Systems with Multiple Delays ... 156
  11.8 Minimum Energy Control of Systems with Single Delay ... 158

12 Controllability of Stochastic Systems with Distributed Delays in Control ... 165
  12.1 Introduction ... 165
  12.2 System Description ... 166
  12.3 Stochastic Relative Controllability ... 170

References ... 173

Chapter 1
Introduction

1.1 Controllability Concept

Control theory is an interdisciplinary branch of engineering and mathematics that deals with influence behavior of dynamical systems. Controllability is one of the fundamental concepts in mathematical control theory. This is a qualitative property of dynamical control systems and is of particular importance in control theory. Systematic study of controllability was started at the beginning of sixties in the last century, when the theory of controllability based on the description in the form of state space for both time-invariant and time-varying linear control systems was worked out.

Roughly speaking, controllability generally means, that it is possible to steer dynamical control system from an arbitrary initial state to an arbitrary final state using the set of admissible controls. It should be mentioned, that in the literature there are many different definitions of controllability, which strongly depend on one hand on a class of dynamical control systems and on the other hand on the form of admissible controls.

In recent years various controllability problems for different types of linear semilinear and nonlinear dynamical systems have been considered in many publications and monographs. Moreover, it should be stressed, that the most literature in this direction has been mainly concerned with different controllability problems for dynamical systems with unconstrained controls and without delays in the state variables or in the controls.

The main purpose of the paper is to present without mathematical proofs a review of recent controllability problems for wide class of dynamical systems. Moreover, it should be pointed out, that exact mathematical descriptions of controllability criteria can be found for example in the following publications [21–58].

1.2 Controllability Significance

Controllability plays an essential role in the development of modern mathematical control theory. There are various important relationships between controllability, stability and stabilizability of linear both finite-dimensional and infinite-dimensional control systems. Controllability is also strongly related with the theory of realization and so called minimal realization and canonical forms for linear time-invariant control systems such as Kalmam canonical form, Jordan canonical form or Luenberger canonical form. It should be mentioned, that for many dynamical systems there exists a formal duality between the concepts of controllability and observability. Moreover, controllability is strongly connected with minimum energy control problem for many classes of linear finite dimensional, infinite dimensional dynamical systems, and delayed systems both deterministic and stochastic.

Therefore, controllability criteria are useful in the following branches of mathematical control theory:

- stabilizability conditions, canonical forms, minimum energy control and minimal realization for positive systems,
- stabilizability conditions, canonical forms, minimum energy control and minimal realization for fractional systems,
- minimum energy control problem for a wide class of stochastic systems with delays in control and state variables,
- duality theorems, canonical forms and minimum energy control for infinite dimensional systems,
- controllability, duality, stabilizability, mathematical modeling and optimal control of quantum systems.

Controllability has many important applications not only in control theory and systems theory, but also in such areas as industrial and chemical process control, reactor control, control of electric bulk power systems, aerospace engineering and recently in quantum systems theory.

Systematic study of controllability was started at the beginning of the sixties in the 20th century, when the theory of controllability based on the description in the form of state space for both time-invariant and time-varying linear control systems was worked out. The extensive list of these publications can be found for example in the monographs [30, 31] or in the survey papers [32, 40, 51].

During last few years quantum dynamical systems have been discussed in many publications. This fact is motivated by possible applications in the theory of quantum informatics [10, 76–78]. Quantum control systems are either defined in finite-dimensional complex space or in the space of linear operators over finite-dimensional complex space. In the first case the quantum states are called state vectors and in the second density operators.

Control system description of a quantum closed system is described by bilinear ordinary differential state equation in the form of Schrödinger equation for state vectors and Liouville [6, 7] equation for density matrices. Therefore, controllability investigations require using special mathematical methods as Lie groups and Lie algebras.

Traditional controllability concept can be extended for so called structural controllability, which may be more reasonable in case of uncertainties [30, 31]. It should be pointed out, that in practice most of system parameter values are difficult to identify and are known only to certain approximations. Thus structural controllability which is independent of a specific value of unknown parameters are of particular interest. Roughly speaking, linear system is said to be structurally controllable if one can find a set of values for the free parameters such that the corresponding system is controllable in the standard sense [30, 31].

Structural controllability of linear control system is strongly related to numerical computations of distance from a given controllable switched linear control system to the nearest an uncontrollable one [30, 31].

First of all let us observe, that from algebraic characterization of controllability and structural controllability immediately follows that controllability is a generic property in the space of matrices defining such systems [30, 31]. Therefore, the set of controllable switched systems is an open and dense subset. Hence, it is important to know how far a controllable linear system is from the nearest uncontrollable linear system. This is very important for linear systems with matrices, whose coefficients are given with some parameter uncertainty.

Explicit bound for the distance between a controllable linear control system to the closed set of uncontrollable switched linear control system can be obtained using special norm defined for the set of matrices and singular value decomposition for controllability matrix [30, 31].

1.3 Delayed Systems

Up to the present time the problem of controllability in continuous and discrete time linear dynamical systems has been extensively investigated in many papers (see e.g. [14, 30–32, 36, 41, 42]). However, this is not true for the nonlinear or semilinear dynamical systems, especially with delays in control and with constrained controls. Only a few papers concern constrained controllability problems for continuous or discrete nonlinear or semilinear dynamical systems with constrained controls [42, 48].

Dynamical systems with distributed [27] delays in control and state variable were also considered. Using some mapping theorems taken from functional analysis and linear approximation methods sufficient conditions for constrained relative and absolute controllability will be derived and proved.

Let us recall that semilinear dynamical control systems with delays may contain different types of delays, both in pure linear and pure nonlinear parts, in the differential state equations. Sufficient conditions for constrained local relative controllability near the origin in a prescribed finite time interval for semilinear dynamical systems with multiple variable point delays or distributed delays in the admissible control and in the state variables, which nonlinear term is continuously differentiable near the origin are presented in [42, 48].

In the above papers it is generally assumed that the values of admissible controls are in a given convex and closed cone with vertex at zero, or in a cone with nonempty interior. The proof of the main result are based on a so called generalized open mapping theorem presented in the paper [68]. Moreover, necessary and sufficient conditions for constrained global relative controllability of an associated linear dynamical system with multiple point delays in control are also discussed.

1.4 Fractional Systems

The development of controllability theory both for continuous-time and discrete-time dynamical systems with fractional derivatives and fractional difference operators has seen considerable advances since the publication of papers [15, 16, 20, 64, 71] and monograph [17]. Although classic mathematical models are still very useful, large dynamical systems prompt the search for more refined mathematical models, which leads to better understanding and approximations of real processes.

The general theory of fractional differential equations and their applications to the field of physics and technique can be found in [17]. This theory formed a very active research topic since provides a natural framework for mathematical modeling of many physical phenomena. In particular, the fast development of this theory has allowed to solve a wide range of problems in mathematical modeling and simulation of certain kinds of dynamical systems in physics and electronics. Fractional derivative techniques provide useful exploratory tools, including the suggestion of new mathematical models and the validation of existing ones.

Mathematical fundamentals of fractional calculus and fractional differential and difference equations are given in the monographs [17], and in the related papers [14–16]. Most of the earliest work on controllability for fractional dynamical systems was related to linear continuous-time or discrete-time systems with limited applications of the real dynamical systems. In addition, the earliest theoretical work concerned time-invariant processes without delays in state variables or in control.

Using the results presented for linear fractional systems and applying linearization method the sufficient conditions for local controllability near the origin are formulated and proved in the paper. Moreover, applying generalized open mapping theorem in Banach spaces [68] and linear semigroup theory in the paper [72] the sufficient conditions for approximation controllability in finite time with conically constrained admissible controls are formulated and proved.

1.5 Positive Systems

In recent years, the theory of positive dynamical systems a natural frame work for mathematical modeling of many real world phenomena, namely in control, biological and medical domains. Positive dynamical systems are of fundamental importance to numerous applications in different areas of science such as economics, biology, sociology and communication [17–19].

Positive dynamical systems both linear and nonlinear are dynamical systems with states, controls and outputs belonging to positive cones in linear spaces. Therefore, in fact positive dynamical systems are nonlinear systems. Among many important developments in control theory over last two decades, control theory of positive dynamical systems [55] has played an essential role.

Controllability, reachability and realization problems for finite dimensional positive both continuous-time and discrete-time dynamical systems were discussed for example in monograph [14] and paper [50], using the results taken directly from the nonlinear functional analysis and especially from the theory of semigroups of bounded operators and general theory of unbounded linear operators.

1.6 Stochastic Systems

Classical control theory generally is based on deterministic approaches. However, uncertainty is a fundamental characteristic of many real dynamical systems. Theory of stochastic dynamical systems are now a well established topic of research, which is still in intensive development and offers many open problems. Important fields of application are economics problems, decision problems, statistical physics, epidemiology, insurance mathematics, reliability theory, risk theory and others methods based on stochastic equations. Stochastic modeling has been widely used to model the phenomena arising in many branches of science and industry such as biology, economics, mechanics, electronics and telecommunications. The inclusion of random effects in differential equations leads to several distinct classes of stochastic equations, for which the solution processes have differentiable or non-differentiable sample paths. Therefore, stochastic differential equations and their controllability require many different method of analysis.

The general theory of stochastic differential equations both finite-dimensional and infinite-dimensional and their applications to the field of physics and technique can be found in the many mathematical monographs and related papers. This theory formed a very active research topic since provides a natural framework for mathematical modeling of many physical phenomena.

Controllability, both for linear or nonlinear stochastic dynamical systems, has recently received the attention of many researchers and has been discussed in several papers and monographs, in which where many different sufficient or necessary and sufficient conditions for stochastic controllability were formulated and proved [8, 41, 55, 56, 61, 62]. However, it should be pointed out that all these results were obtained only for unconstrained admissible controls, finite dimensional state space and without delays in state or control.

Stochastic controllability problems for stochastic infinite-dimensional semilinear impulsive integrodifferential dynamical systems with additive noise and with or without multiple time-varying point delays in the state variables are also discussed in the literature. The proofs of the main results are based on certain theorems taken from the theory of stochastic processes, linearization methods for stochastic dynamical systems, theory of semigroups of linear operators, different fixed-point theorems as Banach, Schauder, Schaefer, or Nussbaum fixed-point theorems and on so-called generalized open mapping theorem presented and proved in many papers (see e.g. [45, 65, 80]).

1.7 Nonlinear and Semilinear Dynamical Systems

The last decades have seen a continually growing interest in controllability theory of dynamical systems. This is clearly related to the wide variety of theoretical results and possible applications. Up to the present time the problem of controllability for continuous-time and discrete-time linear dynamical systems has been extensively investigated in many papers (see e.g. [30–32, 72] for extensive list of references). However, this is not true for the nonlinear dynamical systems especially with different types of delays in control and state variables, and for nonlinear dynamical systems with constrained controls.

Similarly, only a few papers concern constrained controllability problems for continuous or discrete semilinear dynamical systems. It should be pointed out, that in the proofs of controllability results for nonlinear and semilinear dynamical systems linearization methods and generalization of open mapping theorem [3–5] are extensively used. The special case of nonlinear dynamical systems are semilinear systems. Let us recall that semilinear dynamical control systems contain linear and pure nonlinear parts in the differential state equations [3, 35, 66, 69, 80].

1.8 Infinite-Dimensional Systems

Infinite-dimensional dynamical control systems plays a very important role in mathematical control theory. This class consists of both continuous-time systems and discrete-time systems [30–32, 44, 72]. Continuous-time infinite-dimensional systems include for example, a very wide class of so-called distributed parameter systems described by numerous types of partial differential equations defined in bounded or unbounded regions and with different boundary conditions.

1.8 Infinite-Dimensional Systems

For infinite-dimensional dynamical systems it is necessary to distinguish between the notions of approximate and exact controllability [30, 31]. It follows directly from the fact that in infinite-dimensional spaces there exist linear subspaces which are not closed. On the other hand, for nonlinear dynamical systems there exist two fundamental concepts of controllability; namely local controllability and global controllability [30, 31]. Therefore, for nonlinear abstract dynamical systems defined in infinite-dimensional spaces the following four main kinds of controllability are considered: local approximate controllability, global approximate controllability, local exact controllability and global exact controllability [30–32, 44].

Controllability problems for finite-dimensional nonlinear dynamical systems and stochastic dynamical systems have been considered in many publications; see e.g. [30, 31, 44, 55, 56], for review of the literature. However, there exist only a few papers on controllability problems for infinite-dimensional nonlinear systems [65, 66, 80].

Among the fundamental theoretical results, used in the proofs of the main results for nonlinear or semilinear dynamical systems, the most important include:

- generalized open mapping theorem,
- spectral theory of linear unbounded operators,
- linear semigroups theory for bounded linear operators,
- Lie algebras and Lie groups,
- fixed-point theorems such as Banach, Schauder, Schaefer and Nussbaum theorems,
- theory of completely positive trace preserving maps,
- mild solutions of abstract differential and evolution equations in Hilbert and Banach spaces.

1.9 Nonlinear Neutral Impulsive Integrodifferential Evolution Systems in Banach Spaces

In various fields of science and engineering, many problems that are related to linear viscoelasticity, nonlinear elasticity and Newtonian or non-Newtonian fluid mechanics have mathematical models which are described by differential or integral equations or integrodifferential equations. This part of the paper centers around the controllability for dynamical systems described by the integrodifferential models. Such systems are modeled by abstract delay differential equations. In particular abstract neutral differential equations arise in many areas of applied mathematics and, for this reason, this type of equation has been receiving much attention in recent years and they depend on the delays of state and their derivative. Related works of this kind can be found in [9, 65, 70].

The study of differential equations with traditional initial value problem has been extended in several directions. One emerging direction is to consider the impulsive initial conditions. The impulsive initial conditions are combinations of traditional initial value problems and short-term perturbations, whose duration can be negligible in comparison with the duration of the process. Several authors [9, 65, 70] have investigated the impulsive differential equations. In control theory, one of the most important qualitative aspects of a dynamical system is controllability.

As far as the controllability problems associated with finite-dimensional systems modeled by ordinary differential equations are concerned, this theory has been extensively studied during the last decades. In the finite-dimensional context, a system is controllable if and only if the algebraic Kalman rank condition is satisfied. According to this property, when a system is controllable for some time, it is controllable for all time. But this is no longer true in the context of infinite-dimensional systems modeled by partial differential equations. The finite-dimensional controllability of linear, nonlinear and integrodifferential systems has been studied in several publications [27, 33, 36, 42, 48, 63].

The large class of scientific and engineering problems modeled by partial differential and integrodifferential equations can be expressed in various forms of differential and integrodifferential equations in abstract spaces. It is interesting to study the controllability problem for such models in Banach spaces. The controllability problem for first and second order nonlinear functional differential and integrodifferential systems in Banach spaces has been studied by many authors by using semigroup theory, cosine family of operators and various fixed point theorems for nonlinear operators [66, 69] such as Banach theorem, Nussbaum theorem, Schaefer theorem, Schauder theorem, Monch theorem or Sadovski theorem.

In recent years, the theory of impulsive differential equations provides a natural frame work for mathematical modeling of many real world phenomena, namely in control, biological and medical domains. In these models, the investigated simulating processes and phenomena are subjected to certain perturbations hose duration is negligible in comparison with the total duration of the process. Such perturbations can be reasonably well approximated as being instantaneous changes of state, or in the form of impulses. These process tend to be more suitably modeled by impulsive differential equations, which allow for discontinuities in the evolution of the state.

On the other hand, the concept of controllability is of great importance in mathematical control theory. The problem of controllability is to show the existence of a control function, which steers the solution of the system from its initial state to final state, where the initial and final states may vary over the entire space. Many authors have studied the controllability of nonlinear systems with and without impulses, see for instance [3, 4, 27, 32, 34, 36, 63, 69].

In recent years, significant progress has been made in the controllability of linear and nonlinear deterministic systems [5, 59, 65] and the nonlocal initial condition, in many cases, has much better effect in applications then the traditional initial condition. The nonlocal initial value problems can be more useful than the standard initial value problems to describe many physical phenomena of dynamical systems. It should be pointed out, that the study of Volterra-Fredholm integrodifferential equations plays an important role for abstract formulation of many initial, boundary value problems of perturbed differential partial integro-differential equations.

Recently, many authors studied about mixed type integro-differential systems without (or with) delay conditions. Moreover, controllability of impulsive functional differential systems with nonlocal conditions by using the measures of noncompactness and Monch fixed point theorem and some sufficient conditions for controllability were established.

It should be mentioned, that without assuming the compactness of the evolution system the existence, uniqueness and continuous dependence of mild solutions for nonlinear mixed type integrodifferential equations with finite delay and nonlocal conditions has been also established. The results were obtained by using Banach fixed point theorem and semigroup theory. More recently, the existence of mild solutions for the nonlinear mixed type integro-differential functional evolution equations with nonlocal conditions was derived and the results were achieved by using Monch fixed point theorem and fixed point theory.

To the best of our knowledge, up to now no work reported on controllability of impulsive mixed Volterra-Fredholm functional integro-differential evolution differential system with finite delay and nonlocal conditions is an untreated topic in the literature and this fact is the main aim of the present works

1.10 Second Order Impulsive Functional Integrodifferential Systems in Banach Spaces

Second order differential equations arise in many areas of science and technology whenever a relationship involving some continuously changing quantities and their rates of change are known. In particular, second order differential and integrodifferential equations serve as an abstract formulation of many partial integrodifferential equations which arise in problems connected with the transverse motion of an extensible beam, the vibration of hinged bars and many other physical phenomena. So it is quite significant to study the controllability problem for such systems in Banach spaces.

The concept of controllability involves the ability to move a system around in its entire configuration space using only certain admissible manipulations. The exact definition varies slightly within the framework of the type of models. In many cases, it is advantageous to treat the second order abstract differential equations directly rather than to convert them to first order systems. In the proofs of controllability criteria some basic ideas from the theory of cosine families of operators, which is related to the second order equations are often used. Damping may be mathematically modeled as a force synchronous with the velocity of the object but opposite in direction to it. The occurrence of damped second order equations can be found in [59, 65]. The branch of modern applied analysis known as "impulsive" differential equations furnishes a natural framework to mathematically describe some jumping processes.

The theory of impulsive integrodifferential equations and their applications to the field of physics have formed a very active research topic since the theory provides a natural framework for mathematical modeling of many physical phenomena [4, 66]. In spite of the great possibilities for applications, the theory of these equations has been developing rather slowly due to obstacles of theoretical and technical character. The study of the properties of their solutions has been of an ever growing interest.

Recently, most efforts have been focused on the problem of controllability for various kinds of impulsive systems using different approaches [9, 12]. In delay differential equations, the derivative of the unknown function at a certain time is given in terms of the values of the function at previous times. Neutral differential equations arise in many fields and they depend on the delays of state and its derivative. Related works of this kind of equation can be found in [13, 63]. For the fundamental solution of second order evolution system, one can refer the paper [60].

1.11 Switched Systems

The last decades have seen a continually growing interest in controllability theory of hybrid dynamical systems and their special case named switched dynamical systems. In the literature there have been a lot of papers for controllability both continuous-time and discrete-time switched systems [1, 2, 11, 19, 67, 73–75, 79]. Switched systems deserve investigation for theoretical interest as well as for practical applications. Switching system structure is an essential feature of many engineering control applications such as power systems and power electronics. From a theoretical point of view switched linear system consists of several linear subsystems and a rule that organize switching among them.

Hybrid systems which are capable of exhibiting simultaneously several kinds of dynamic behavior in different parts of the system (e.g., continuous-time dynamics, discrete-time dynamics, jump phenomena, logic commands) are of great current interest (see, e.g., [11, 75]). Examples of such systems include the Multiple-Models, Switching and Tuning paradigm from adaptive control, Hybrid Control Systems, and a plethora of techniques that arise in Event Driven Systems are typical examples of such systems of varying degrees of complexity. Moreover, hybrid systems include computer disk drives, transmission an stepper motors, constrained robotic systems, intelligent vehicle/highway systems, sampled-data systems, discrete event systems, and many other types of dynamical systems.

Switched linear systems are hybrid systems that consist of several linear subsystems and a rule of switching among them. Switched linear systems provide a framework which bridges the linear systems and the complex and/or uncertain systems. On one hand, switching among linear systems may produce complex system behaviors such as chaos and multiple limit cycles. On the other hand, switched linear systems are relatively easy to handle as many powerful tools from linear and multilinear analysis are available to cope with these systems.

1.11 Switched Systems

Moreover, the study of switched linear systems provides additional insights into some long-standing and sophisticated problems, such as intelligent control, adaptive control, and robust analysis and control. Theoretical examination of switched linear systems are academically more challenging due to their rich, diverse, and complex dynamics. Switching makes those systems much more complicated than standard-time invariant or even time-varying systems. Many more complicated behaviors/dynamics and fundamentally new properties, which standard systems do not have, have been demonstrated on switched linear systems. From the point of view of control system design, switching brings an additional degree of freedom in control system design. Switching laws, in addition to control laws, may be utilized to manipulate switched systems to achieve a better performance of a system. This can be seen as an added advantage for control design to attain certain control purposes like stabilizability or controllability.

For controllability analysis of switched linear control systems, a much more difficult situation arises since both the control input and the switching rule are design variables to be determined. Thus, the interaction between them is very important from controllability point of view. Moreover, it should be mentioned that for switched linear discrete-time control system in general case the controllable set is not a subspace but a countable union of subspaces. For switched linear continuous-time control system, in general case the controllable set is an uncountable union of subspaces.

Controllability problems for different types of dynamical systems require the application of numerous mathematical concepts and methods taken directly from differential geometry, functional analysis, topology, matrix analysis and theory of ordinary and partial differential equations and theory of difference equations. The state-space models of dynamical systems provides a robust and universal method for studying controllability of various classes of systems.

Finally, it should be stressed, that there are numerous open problems for controllability concepts for special types of dynamical systems. For example, it should be pointed out, that up to present time the most literature on controllability problems has been mainly concerned with unconstrained controls and without delays in the state variables or in the controls.

1.12 Quantum Dynamical Systems

Fast recent development of quantum information field in both theory and experiments caused increased interest in new methods of quantum systems control. Various models for open-loop and closed-loop control scenarios for quantum systems have been developed in recent years [10, 76–78].

Quantum systems can be classified according to their interaction with the environment. If a quantum system exchange neither information nor energy with its environment it is called closed and its time evolution is described completely by a Hamiltonian and its respective unitary operator. On the other hand if the exchange of information or energy occurs, the system is called open.

Due to the destructive nature of quantum measurement in some models one has to be constrained to open-loop control of a quantum system. This fact means that during the time evolution of the quantum system it is physically impossible to extract any information about the state of the system.

In the simplest case open-loop control of the closed quantum system is described by the bi-linear model. In this case the differential equation of the evolution is described by the sum of the drift Hamiltonian and the control Hamiltonians. The parameters of the control Hamiltonians may be constrained in various ways due to physical constraints of the system. Many quantum systems can be only controlled locally, which means that control Hamiltonians act only on one of the Hilbert spaces that constitute larger tensor product Hilbert space of the system.

The control constrained to local operations is of a great interest in various applications, especially in quantum computation and spin graph systems. Other possible constraints, such as constrained energy or constrained frequency, are possible. They are very important in the scope of optimal control of quantum systems.

In the most generic case open quantum systems are not controllable with coherent, unitary control due to the fact that the action of the generic completely positive trace preserving map cannot be reversed unitarily. For example Markovian dynamics of finite-dimensional open quantum system is not coherently controllable. However, many schemes of incoherent control of open quantum systems have been described. Some of these schemes are based on the technique known as quantum error correcting codes. In incoherent control schemes quantum unitary evolution together with quantum measurements is used to drive the system to the desired state even if quantum noise is present in the system.

Chapter 2
Controllability and Minimum Energy Control of Linear Finite Dimensional Systems

2.1 Introduction

Stability, controllability and observability are one of the fundamental concepts in modern mathematical control theory. They are qualitative properties of control systems and are of particular importance in control theory. Systematic study of controllability and observability was started at the beginning of sixties, when the theory of controllability and observability based on the description in the form of state space for both time-invariant and time-varying linear control systems was worked out. The concept of stability is extremely important, because almost every workable control system is designed to be stable. If a control system is not stable, it is usually of no use in practice.

Many dynamical systems are such that the control does not affect the complete state of the dynamical system but only a part of it. On the other hand, very often in real industrial processes it is possible to observe only a certain part of the complete state of the dynamical system. Therefore, it is very important to determine whether or not control and observation of the complete state of the dynamical system are possible. Roughly speaking, controllability generally means, that it is possible to steer dynamical system from an arbitrary initial state to an arbitrary final state using the set of admissible controls.

On the other hand observability means, that it is possible to recover uniquely the initial state of the dynamical system from a knowledge of the input and output. Stability controllability and observability play an essential role in the development of the modern mathematical control theory. There are important relationships between controllability, observability and stabilizability of linear control systems. Controllability and observability are also strongly connected with the theory of minimal realization of linear time-invariant control systems. Moreover, it should be pointed out that there exists a formal duality between the concepts of controllability and observability.

In the literature there are many different definitions of stability, controllability and observability which depend on the type of dynamical control system. The main purpose of this article is to present a compact review over the existing stability, controllability and observability results mainly for linear continuous-time and time-invariant control systems. It should be pointed out that for linear control systems stability, controllability and observability conditions have pure algebraic forms and are rather easily computable. These conditions require verification location of the roots of a characteristic polynomial and of the rank conditions for suitable defined constant controllability and observability matrices.

The chapter is organized as follows: Sect. 2.2 contains systems descriptions and fundamental results concerning the solution of the most popular linear continuous-time control models with constant coefficients. Section 2.3 is devoted to a study of different kinds of stability. Section 2.4 presents fundamental definitions of controllability and necessary and sufficient conditions for different kinds of controllability. Section 2.2 contains fundamental definition of observability and necessary and sufficient conditions for observability. Finally, in concluding remarks and comments concerning possible extensions for more general cases are presented. Since the article should be limited to a reasonable size, it is impossible to give a full survey on the subject. In consequence, only selected fundamental results without proofs are presented.

2.2 Mathematical Model

In the theory of linear time-invariant dynamical control systems the most popular and the most frequently used mathematical model is given by the following differential state equation and algebraic output equations

$$x'(t) = Ax(t) + Bu(t) \tag{2.1}$$

$$y(t) = Cx(t) \tag{2.2}$$

where $x(t) \in R^n$ is a state vector, $u(t) \in R^m$ is an input vector, $y(t) \in R^p$ is an output vector, A, B, and C are real matrices of appropriate dimensions.

It is well known that for a given initial state $x(0) \in R^n$ and control $u(t) \in R^m$, $t \geq 0$, there exist unique solution $x(t; x(0), u) \in R^n$ of the state Eq. (2.1) of the following form

$$x(t; x(0), u) = \exp(At)x(0) + \int_0^t \exp(A(t-s))Bu(s)ds$$

2.2 Mathematical Model

Let P be an $n \times n$ constant nonsingular transformation matrix and let us define the equivalence transformation $z(t) = Px(t)$. Then the state Eq. (2.1) and output Eq. (2.2) becomes

$$z'(t) = Jz(t) + Gu(t) \tag{2.3}$$

$$y(t) = Hz(t) \tag{2.4}$$

where

$$J = PAP^{-1}, G = PB \quad \text{and} \quad H = CP^{-1}.$$

Dynamical systems (2.1)–(2.4) are said to be equivalent and many of their properties are invariant under the equivalence transformations.

Among different equivalence transformations special attention should be paid on the transformation which leads to the so called Jordan canonical form of the dynamical system. If the matrix J is in the Jordan canonical form, then Eqs. (2.3) and (2.4) are said to be in a Jordan canonical form. It should be stressed, that every dynamical system (2.1) and (2.2) has an equivalent Jordan canonical form.

2.3 Controllability Conditions

Now, let us recall the most popular and frequently used fundamental definition of controllability for linear control systems with constant coefficients.

Definition 2.1 Dynamical system (2.1) is said to be controllable if for every initial condition $x(0)$ and every vector $x^1 \in R^n$, there exist a finite time t_1 and control $u(t) \in R^m$, $t \in [0, t_1]$, such that $x(t_1; x(0), u) = x^1$.

This definition requires only that any initial state $x(0)$ can be steered to any final state x^1. It should be pointed out, that trajectory of the system in the time interval $[0, t_1]$ is not specified. Furthermore, there is no constraints imposed on the control.

In order to formulate easily computable algebraic controllability criteria let us introduce the so called controllability matrix W defined as follows.

$$W = \begin{bmatrix} B, AB, A^2B, \ldots, A^{n-1}B \end{bmatrix}.$$

Controllability matrix W is an $n \times nm$-dimensional constant matrix and depends only on system parameters.

Theorem 2.1 *Dynamical system* (2.1) *is controllable if and only if*

$$\text{rank } W = n$$

Corollary 2.1 *Dynamical system* (2.1) *is controllable if and only if the $n \times n$-dimensional symmetric matrix WW^T is nonsingular.*

Since the controllability matrix W does not depend on time t_1, then from Theorem 2.1 and Corollary 2.1 it directly follows, that in fact controllability of dynamical system does not depend on the length of control interval.

Let us observe, that in many cases in order to check controllability, it is not necessary to calculate the controllability matrix W but only a matrix with a smaller number of columns. It depends on the rank of the matrix B and the degree of the minimal polynomial of the matrix A, where the minimal polynomial is the polynomial of the lowest degree which annihilates matrix A. This is based on the following Corollary.

Corollary 2.2 *Let rank $B = r$, and q is the degree of the minimal polynomial of the matrix A. Then dynamical system (2.1) is controllable if and only if*

$$rank\left[B, AB, A^2B, \ldots, A^{n-k}B\right] = n$$

where the integer $k \leq min(n - r, q - 1)$.

In the case when the eigenvalues of the matrix A, s_i, $i = 1,2,3,\ldots,n$ are known, we can check controllability using the following Corollary.

Corollary 2.3 *Dynamical system (2.1) is controllable if and only if*

$$rank[s_iI - A|B] = n \quad \text{for all } s_i, \quad i = 1, 2, 3, \ldots, n$$

Suppose that the dynamical system (2.1) is controllable, then the dynamical system remains controllable after the equivalence transformation. This is natural and intuitively clear because an equivalence transformation changes only the basis of the state space. Therefore, we have the following Corollary.

Corollary 2.4 *Controllability is invariant under any equivalence transformation*

$$z(t) = Px(t).$$

Since controllability of a dynamical system is preserved under any equivalence transformation, then it is possible to obtain a simpler controllability criterion by transforming the differential state Eq. (2.1) into a special form (2.3).

It should be pointed out, that if we transform dynamical system (2.1) into Jordan canonical form, then controllability can be determined very easily, almost by inspection.

2.4 Stabilizability

It is well known, that the controllability concept for dynamical system (2.1) is strongly related to its stabilizability by the linear static state feedback of the following form

2.4 Stabilizability

$$u(t) = Kx(t) + v(t) \quad (2.5)$$

where $v(t) \in R^m$ is a new control, K is $m \times n$-dimensional constant state feedback matrix.

Introducing the linear static state feedback given by equality (2.5) we directly obtain the linear differential state equation for the feedback linear dynamical system of the following form

$$x'(t) = (A + BK)x(t) + Bv(t) \quad (2.6)$$

which is characterized by the pair of constant matrices $(A + BK, B)$.

An interesting result is the equivalence between controllability of the dynamical systems (2.1) and (2.6), explained in the following Corollary.

Corollary 2.5 *Dynamical system (2.1) is controllable if and only if for arbitrary matrix K the dynamical system (2.6) is controllable.*

From Corollary 2.5 it follows that under the controllability assumption we can arbitrarily form the spectrum of the dynamical system (2.1) by the introduction of suitable defined linear static state feedback (2.5). Hence, we have the following result.

Theorem 2.2 *The pair of matrices (A, B) represents the controllable dynamical system (2.1) if and only if for each set Λ consisting of n complex numbers and symmetric with respect to real axis, there exists constant state feedback matrix K such that the spectrum of the matrix $(A + BK)$ is equal to the set Λ.*

Practically, in the design of the dynamical system, sometimes it is required only to change unstable eigenvalues, i.e. the eigenvalues with nonnegative real parts into stable eigenvalues, with negative real parts. This is called stabilization of the dynamical system (2.1). Therefore, we have the following formal definition of stabilizability.

Definition 2.2 The dynamical system (2.1) is said to be stabilizable if there exists constant static state feedback matrix K such that the spectrum of the matrix $(A + BK)$ entirely lies in the left-hand side of the complex plane.

Let $Re(s_j) \geq 0$, for $j = 1,2,3,...,q \leq n$, i.e., s_j are unstable eigenvalues of the dynamical system (2.1). An immediate relation between controllability and stabilizability of the dynamical system (2.1) gives the following Theorem.

Theorem 2.3 *The dynamical system (2.1) is stabilizable if and only if all its unstable modes are controllable i.e.,*

$$\text{rank}\begin{bmatrix}s_j I - A | B\end{bmatrix} = n \quad \text{for } j = 1, 2, 3, \ldots, q$$

Comparing Theorem 2.3 and Corollary 2.3 we see, that controllability of the dynamical system (2.1) always implies its stabilizability, but the converse statement is not always true. Therefore, the stabilizability concept is essentially weaker than the controllability.

2.5 Output Controllability

Similar to the state controllability of a dynamical control system, it is possible to define the so called output controllability for the output vector y(t) of a dynamical system. Although these two concepts are quite similar, it should be mentioned that the state controllability is a property of the differential state Eq. (2.1), whereas the output controllability is a property both of the state Eq. (2.1) and algebraic output Eq. (2.2).

Definition 2.3 Dynamical system (2.1), (2.2) is said to be output controllable if for every $y(0)$ and every vector $y^1 \in R^p$, there exist a finite time t_1 and control $u^1(t) \in R^m$, that transfers the output from $y(0)$ to $y^1 = y(t_1)$.

Theorem 2.4 *Dynamical system (2.1), (2.2) is output controllable if and only if*

$$rank\left[CB, CAB, CA^2B, \ldots, CA^{n-1}B\right] = p$$

It should be pointed out, that the state controllability is defined only for the linear differential state Eq. (2.1), whereas the output controllability is defined for the input-output description, i.e. it depends also on the linear algebraic output Eq. (2.2). Therefore, these two concepts are not necessarily related.

If the control system is output controllable, its output can be transferred to any desired vector at certain instant of time. A related problem is whether it is possible to steer the output following a preassigned curve over any interval of time. A control system whose output can be steered along the arbitrary given curve over any interval of time is said to be output function controllable or functional reproducible.

2.6 Controllability With Constrained Controls

In practice admissible controls are required to satisfy additional constraints. Let $U \subset R^m$ be an arbitrary set and let the symbol $M(U)$ denotes the set of admissible controls i.e., the set of controls $u(t) \in U$ for $t \in [0, \infty)$.

Definition 2.4 The dynamical system (2.1) is said to be U-controllable to zero if for any initial state $x(0) \in R^n$, there exist a finite time $t_1 < \infty$ and an admissible control $u(t) \in M(U)$, $t \in [0, t_1]$, such that $x(t_1; x(0), u) = x^1$.

Definition 2.5. The dynamical system (2.1) is said to be U-controllable from zero if for any final state $x^1 \in R^n$, there exist a finite time $t_1 < \infty$ and an admissible control $u(t) \in M(U)$, $t \in [0, t_1]$, such that $x(t_1; 0, u) = x^1$.

Definition 2.6 The dynamical system (2.1) is said to be U-controllable if for any initial state $x(0) \in R^n$, and any final state $x^1 \in R^n$, there exist a finite time $t_1 < \infty$ and an admissible control $u(t) \in M(U)$, $t \in [0, t_1]$, such that $x(t_1; x(0), u) = x^1$.

2.6 Controllability With Constrained Controls

Generally, for arbitrary set U it is rather difficult to give easily computable criteria for constrained controllability. However, for certain special cases of the set U it is possible to formulate and prove algebraic constrained controllability conditions.

Theorem 2.5 *The dynamical system (2.1) is U-controllable to zero if and only if all the following conditions are satisfied simultaneously*:

(1) *there exists $w \in U$ such that $Bw = 0$*,
(2) *the convex hull $CH(U)$ of the set U has nonempty interior in the space R^m*,
(3) *rank$[B, AB, A^2B,...,A^{n-1}B] = n$*,
(4) *there is no real eigenvector $v \in R^n$ of the matrix A^{tr} satisfying $v^{tr}Bw \leq 0$ for all $w \in U$*,
(5) *no eigenvalue of the matrix A has a positive real part*.

In the special case for the single input system, i.e. $m = 1$, Theorem 2.5 reduces to the following Corollary.

Corollary 2.6 *Suppose that $m = 1$ and $U = [0,1]$. Then the dynamical system (2.1) is U-controllable to zero if and only if it is controllable without any constraints*, i.e.

$$rank\left[B, AB, A^2B, \ldots, A^{n-1}B\right] = n,$$

and matrix A has only complex eigenvalues.

Theorem 2.6 *Suppose the set U is a cone with vertex at zero and a nonempty interior in the space R^m. Then the dynamical system (2.1) is U-controllable from zero if and only if*

(i) *rank$\left[B, AB, A^2B, \ldots, A^{n-1}B\right] = n$*,
(ii) *there is no real eigenvector $v \in R^n$ of the matrix A^{tr} satisfying $v^{tr}Bw \leq 0$ for all $w \in U$*,

For the single input system, i.e. $m = 1$, Theorem 2.6 reduces to the following Corollary.

Corollary 2.7 *Suppose that $m = 1$ and $U = [0,1]$. Then the dynamical system (2.1) is U-controllable from zero if and only if it is controllable without any constraints*, i.e.

$$rank\left[B, AB, A^2B, \ldots, A^{n-1}B\right] = n,$$

and matrix A has only complex eigenvalues.

2.7 Controllability After Introducing Sampling

We consider now the case in which the control u is piecewise constant, i.e. the control u changes value only at a discrete instant of time. Inputs of this type occur in sampled-data systems or in systems, in which digital computers are used to generate control u. A piecewise-constant control u is often generated by a sampler and a filter, called zero-order hold. In this case

$$u(t) = u(k) \quad \text{for } kT \leq t < (k+1)T \quad k = 0, 1, 2, \ldots$$

where T is a positive constant, called the sampling period. The discrete times 0, T, $2T, \ldots$ are called sampling instant.

The behavior at sampling instant $0, T, 2T, \ldots$ of the dynamical system (2.1), (2.2) with the piecewise-constant inputs are described by the discrete-time state and output equations

$$x(k+1) = Ex(k) + Fu(k) \tag{2.7}$$

$$y(k) = Cx(k) \tag{2.8}$$

where the matrices E and F are computed as follows

$$E = exp(AT)$$

$$F = \left(\int_0^T exp(As)ds\right) B$$

If the dynamical system (2.1) is controllable, it is of interest to study whether the system remains controllable after the introducing of sampling or equivalently, whether the discrete-time dynamical system (2.7) is controllable.

Theorem 2.7 *Assume that the dynamical system (2.1) is controllable. Then the discrete-time system (2.7) is also controllable if*

$$Im(s_i - s_j) \neq 2\pi q/T$$
for $q = \ldots -2, -1, +1, +2, \ldots$ *whenever* $Re(s_i - s_j) = 0$

Corollary 2.8 *For the single-input case, i.e., for m =1, the condition stated in the Theorem 2.7 is necessary as well.*

Corollary 2.9 *If the dynamical system (2.1) has only real eigenvalues, i.e., $Ims_i = 0$, for $i = 1,2,3,\ldots,r$, then the discrete-time system (2.7) is always controllable.*

Corollary 2.10 *Dynamical system (2.1) with single input and only real eigenvalues is controllable if and only if the discrete-time dynamical system (2.7) is controllable.*

2.8 Perturbations of Controllable Dynamical Systems

In practice the fundamental problem is the question, which bounded perturbation of the parameters of the dynamical linear system (2.1) preserve controllability property.

2.8 Perturbations of Controllable Dynamical Systems

Theorem 2.8 *Suppose the dynamical system (2.1) is controllable. Then there exists $\varepsilon > 0$ such that if*

$$\|A - F\| + \|B - G\| < \varepsilon \tag{2.9}$$

then the dynamical system of the form

$$z'(t) = Mz(t) + Nu(t) \quad z(t) \in R^n$$

is also controllable for any constant matrices H and G of appropriate dimensions, satisfying the inequality (2.9).

Theorem 2.8 can be used in the investigations of the topological properties of the set of controllable systems.

Corollary 2.11 *The set of dynamical systems which are controllable is open and dense in the space of all dynamical system of the form (2.1).*

Corollary 2.11 is of great practical importance. It states that almost all dynamical systems of the form (2.1) are controllable. Therefore, controllability is the so called "generic" property of the dynamical system (2.1). Intuitively this means, that almost all dynamical systems are controllable and, moreover, that for almost all dynamical systems there exist open neighborhoods containing entirely only controllable dynamical systems.

Moreover, Corollary 2.11 enable us to define the so called controllability margin. The controllability margin for dynamical system (2.1) is defined as the distance between the given dynamical system and the nearest non-controllable dynamical system. It is obvious that dynamical system (2.1) which is not controllable has the controllability margin equal to zero.

2.9 Minimum Energy Control

Minimum energy control problem is strongly related to controllability problem. For controllable dynamical system (2.1) there exists generally many different controls which steer the system from a given initial state $x(0)$ to the final desired state x^1 at time $t_1 > 0$. Therefore, we may look for the control which is an optimal in the sense of the following performance index.

$$J(u) = \int_0^{t_1} \|u(t)\|_Q^2 dt$$

where

$$\|u(t)\|_Q^2 = u^{tr}(t)Qu(t)$$

and Q is an $m \times m$-dimensional constant symmetric and positive definite weighting matrix.

The performance index $J(u)$ defines the control energy in the time interval $[0, t_1]$ with the weight determined by the weighting matrix Q. The control u which minimizes the performance index J(u) is called the minimum energy control. It should be mentioned, that the performance index $J(u)$ is a special case of the general quadratic performance index, and hence the existence of a minimizing control function is assured.

Therefore, the minimum energy control problem can be formulated as follows: for a given arbitrary initial state $x(0)$, arbitrary final state x^1, and finite time $t_1 > 0$, find an optimal control $u(t)$, $t \in [0, t_1]$, which transfers the state $x(0)$ to x^1 at time t_1 and minimizes the performance index $J(u)$.

In order to solve the minimum control problem and to present it in a readable compact form, let us introduce the following notation:

$$W_Q = \int_0^{t_1} \exp(At) B Q^{-1} B^{tr} (\exp(At))^{tr} dt,$$

W_Q is constant $n \times n$-dimensional symmetric matrix

$$u^0(t) = Q^{-1} B^{tr} [\exp(A(t_1 - t))]^{tr} W_Q^{-1} [x^1 - exp(At_1) x(0)] \qquad (2.10)$$

Exact analytical solution of the minimum energy control problem for dynamical system (2.1) is given by the following Theorem.

Theorem 2.9 *Let $u^1(t)$, $t \in [0, t_1]$ be any control that transfers initial state $x(0)$ to final state x^1 at time t_1, and let $u^0(t)$, $t \in [0, t_1]$ be the control defined by (2.10). Then the control $u^0(t)$ transfers the initial state $x(0)$ to a final state x^1 at time t_1 and*

$$J(u^1) \geq J(u^0)$$

Moreover, the minimum value of the performance index corresponding to the optimal control u^0 is given by the following formula

$$J(u^0) = [x^1 - exp(At_1) x(0)]^{tr} W_Q^{-1} [x^1 - exp(At_1) x(0)]$$

It should be pointed out the Theorem 2.9 is proved under the following general assumptions:

1. dynamical system is linear,
2. there are no constraints in control,
3. there are no constraints posed on state variable $x(t)$,
4. dynamical system is controllable,
5. performance index $J(u)$ does not contain the state variable $x(t)$,
6. performance index is a quadratic with respect to control $u(t)$.

Moreover, it should be mentioned, that Theorem 2.9 can be proved by using only the fundamental properties of the norm and scalar product in the Hilbert spaces R^n and $L^2([0, t_1], R^m)$, without applying rather complicated results taken from the mathematical optimal control theory. This is possible only under the six general assumptions mentioned before.

The performance index $J(u)$ can also be used in determining the so called controllability measures, which characterize qualitatively the dynamical system (2.1). In the literature there are many different measures of controllability, generally depending on the eigenvalues of the matrix W_Q^{-1}.

2.10 Linear Time-Varying Systems

Now, let us consider linear time varying dynamical control systems, for which mathematical model is given by the following differential state equation with time varying parameters

$$x'(t) = A(t)x(t) + B(t)u(t) \quad t \in [t_0, t_1] \tag{2.11}$$

where similarly as before $x(t) \in R^n$ is a state vector, $u(t) \in R^m$ is an input vector, $A(t)$, and $B(t)$ are matrices of appropriate dimensions with continuous-time elements.

It is well known that for a given initial state $x(t_0) \in R^n$ and control $u(t) \in R^m$, $t \geq t_0$, there exist unique solution $x(t; x(t_0), u) \in R^n$ of the state equation of the following form

$$x(t; x(t_0), u) = H(t, t_0)x(t_0) + \int_{t_0}^{t} H(t, s)B(s)u(s)ds$$

where $H(t, s)$ is fundamental matrix solution for equation

Now, let us recall the most popular and frequently used fundamental definition of controllability for linear time-varying control systems with varying coefficients.

Definition 2.7 Dynamical system (2.11) is said to be controllable in a given time interval $[t_0, t_1]$ for every initial condition $x(t_0)$ and every vector $x^1 \in R^n$, given in finite time t_1 there exist control $u(t) \in R^m$, $t \in [t_0, t_1]$, such that $x(t_1; x(t_1), u) = x^1$.

Similarly as for time invariant case this definition requires only that any initial state $x(t_0)$ can be steered to any final state x^1 at time t_1. The trajectory of the system is not specified. Furthermore, there is no constraints imposed on the control.

In order to formulate easily computable algebraic controllability criteria let us introduce the so called controllability matrix $W(t_0, t_1)$ defined as follows.

$$W(t_0, t_1) = \int_0^{t_1} H(t_1, t)B(t)B^{tr}(t)H(t_1, t)^{tr} dt,$$

Controllability matrix $W(t_0, t_1)$ is an $n \times n$-dimensional constant matrix and depends only on system parameters.

Theorem 2.10 *Dynamical system (2.11) is controllable if and only if*

$$\text{rank } W(t_0, t_1) = n$$

Corollary 2.12 *Dynamical system (2.11) is controllable if and only if the $n \times n$-dimensional symmetric matrix $W(t_0, t_1)W(t_0, t_1)^T$ is nonsingular.*

Minimum energy control problem for time-varying system is defined similarly as for time-invariant case and of course is also strongly related to controllability problem.

Therefore, we may look for the control which steer the system from a given initial state $x(t_0)$ to the final desired state x^1 at time t_1 and is an optimal in the sense of the following performance index.

$$J(u) = \int_0^{t_1} \|u(t)\|_Q^2 dt$$

where

$$\|u(t)\|_Q^2 = u^{tr}(t)Qu(t)$$

and Q is an $m \times m$-dimensional constant symmetric and positive definite weighting matrix.

The performance index $J(u)$ defines the control energy in the time interval $[t_0, t_1]$ with the weight determined by the weighting matrix Q. The control u which minimizes the performance index $J(u)$ is called the minimum energy control. It should be mentioned, that the performance index $J(u)$ is a special case of the general quadratic performance index, and hence the existence of a minimizing control function is assumed.

Therefore, the minimum energy control problem can be formulated as follows: for a given arbitrary initial state $x(t_0)$, arbitrary final state x^1, and finite time t_1, find an optimal control $u(t)$, $t \in [t_0, t_1]$, which transfers the state $x(t_0)$ to x^1 at time t_1 and minimizes the performance index $J(u)$.

In order to solve the minimum control problem and to present it in a readable compact form, let us introduce the following notation:

$$W_Q(t_0, t_1) = \int_0^{t_1} W(t_1, t)B(t)Q^{-1}B^{tr}(t)W(t_1, t)^{tr} dt$$

$W_Q(t_0, t_1)$ is constant $n \times n$-dimensional symmetric matrix

It should be pointed out, that since matrix Q is nonsingular, then matrix $W_Q(t_0, t_1)$ is nonsingular if and only if matrix $W(t_0, t_1)$ is nonsingular.

2.10 Linear Time-Varying Systems

Let us define the admissible control of the following form

$$u^0(t) = Q^{-1} B(t)^{tr} [H(t_1,t)]^{tr} W_Q^{-1} [x^1 - H(t_1,t_0) x(t_0)] \qquad (2.12)$$

Exact analytical solution of the minimum energy control problem for dynamical system (2.11) is given by the following Theorem.

Theorem 2.11 *Let $u^1(t)$, $t \in [t_0, t_1]$ be any control that transfers initial state $x(0)$ to final state x^1 at time t_1, and let $u^0(t)$, $t \in [t_0, t_1]$ be the control defined by (2.12). Then the control $u^0(t)$ transfers the initial state $x(t_0)$ to a final state x^1 at time t_1 and*

$$J(u^1) \geq J(u^0)$$

Moreover, the minimum value of the performance index corresponding to the optimal control u^0 is given by the following formula

$$J(u^0) = [x^1 - H(t_1,t)x(t_0)]^{tr} W_Q^{-1} [x^1 - H(t_1,t)x(t_0)]$$

It should be pointed out the Theorem 2.11 is proved under the following general assumptions:

1. dynamical system is linear,
2. there are no any constraints in control,
3. there are no constraints posed on state variable $x(t)$,
4. dynamical system is controllable,
5. performance index $J(u)$ does not contain the state variable $x(t)$,
6. performance index is a quadratic with respect to control $u(t)$.

Moreover, it should be mentioned, that Theorem 2.11 can be proved by using only the fundamental properties of the norm and scalar product in the Hilbert spaces R^n and $L^2([0, t_1], R^m)$, without applying rather complicated results taken from the mathematical optimal control theory. This is possible only under the six general assumptions mentioned before.

The performance index $J(u)$ can also be used in determining the so called controllability measures, which characterize qualitatively the dynamical system (2.1). In the literature there are many different measures of controllability, generally depending on the eigenvalues of the matrix W_Q^{-1}.

Chapter 3
Controllability of Linear Systems with Delays in State and Control

3.1 Introduction

In the theory of dynamical systems with delays we may consider many different kinds of delays both in state variables or in admissible controls, e.g., distributed delays, point time-variable delays, point constant delays. Thus, in dynamical systems with delays it is necessary to introduce two kinds of state, namely: complete state and instantaneous state. Hence, we have two types of controllability: absolute controllability for complete states and relative controllability for instantaneous states.

The main purpose of this chapter is to study the relative controllability of linear finite delay dynamical systems containing both multiple lumped time varying delays and distributed delays in the state variables and multiple lumped time varying delays in the admissible controls.

3.2 System Description

Let us consider infinite delay dynamical systems containing both multiple lumped time varying delays and distributed delays in the state variables and multiple lumped time varying delays in admissible controls described by the following differential state equation

$$x'(t) = L(t, x_t) + \int_{t_0-h}^{t_0} A(s)x(t+s)ds + \sum_{i=0}^{i=M} B_i(t)u(v_i(t)) \quad t \in [t_0 - h, t_1] \quad (3.1)$$

with given initial condition $x(t) = \phi(t)$, for $t \in (t_0-h, t_0]$.

In Eq. (3.1) symbol $L(t, \phi)$ denotes an operator which is continuous in t, linear in ϕ and given by following equality

$$L(t,\phi) = \sum_{i=0}^{i=M} A_i(t)\phi(v_i(t)) \qquad (3.2)$$

The strictly increasing and twice continuously differentiable functions $v_i(t)$: $[t_0, t_1] \to R$, $i = 0, 1, 2,\ldots, M$, represent deviating arguments in the admissible controls and in the state variables, i.e. $v_i(t) = t - h_i(t)$, where $h_i(t)$ are lumped time varying delays for $i = 0, 1, 2,\ldots M$. Moreover, $v_i(t) \leq t$ for $t \in [t_0,t_1]$, and $i = 0, 1, 2, 3, \ldots, M$.

Let us introduce the time-lead functions $r_i(t)$: $[v_i(t_0), v_i(t_1)] \to [t_0, t_1]$, $i = 0, 1, 2, 3,\ldots, M$, such that $r_i(v_i(t)) = t$ for $t \in [t_0, t_1]$. Furthermore, only for simplicity and compactness of notation let us assume that $v_0(t) = t$ and for a given t_1 the functions $v_i(t)$ satisfy the following inequalities [31, 32].

$$\begin{aligned} h = v_M(t_1) \leq v_{M-1}(t_1) \leq \ldots \leq v_{m+1}(t_1) \leq t_0 \\ = v_m(t_1) < v_{m-1}(t_1) \leq \ldots \leq v_1(t_1) \leq v_0(t_1) = t_1 \end{aligned} \qquad (3.3)$$

In Eq. (3.1) $A_i(t)$, $i = 0, 1, 2,\ldots, M$ are continuous $n \times n$ dimensional matrices, and $A(s)$ is $n \times n$ dimensional matrix whose elements are square integrable functions on $(t_0-h, t_0]$.

The matrix functions $B_i(t)$, $i = 0, 1, 2,\ldots, M$ are $n \times p$ dimensional and continuous in t. The admissible controls $u \in L_2 ([t_0-h, t_1], R^p)$.

Let $h \geq t_0-v_M(t_0) > 0$ be given. For a given function x: $[t_0-h, t_1] \to R^n$ and $t \in [t_0, t_1]$, the symbol x_t denotes as usually the function on $[-h,0]$ defined by $x_t(s) = x(t + s)$ for $s \in [-h, 0]$.

Similarly, for a given control function u: $[v_M(t_0), t_1] \to R^p$, and $t \in [t_0, t_1]$, the symbol u_t denotes the function on $[v_M(t), t)$ defined by the equality $u_t(s) = u(t + s)$ for $s \in [v_M(t), t)$. For example, u_{t_0} is the initial control function defined on time interval $[v_M(t_0), t_0)$.

The function η: $[-h, 0] \to (0, \infty)$ is Lebesque integrable on $[-h, 0]$, positive and nondecreasing. Moreover, let $B([-h, 0], R^n)$ be the Banach space of functions, which are continuous and bounded on $[-h, 0]$ and such that

$$|\phi| = \sup\{|\phi(s)| : s \in [-h, 0]\} + \int_{-h}^{0} \eta(s)|\phi(s)|ds < \infty$$

Let $n \times n$ matrix $X(t, s)$ satisfy the following equations

$$\begin{aligned} \partial/\partial t(X(t,s)) &= L(t, X_t(\bullet, s)) & t \geq s \\ X(t,s) &= 0 & s-h \leq t \leq s \\ X(t,s) &= I & t = s \end{aligned}$$

3.2 System Description

where

$$X_t(\bullet, s)(\tau) = X(t+\tau, s), \quad \text{for } \tau \in [-h, 0]$$

Then, the solution of the Eq. (3.1) is given by

$$x(t) = x(t; t_0, \phi) + \int_{t_0}^{t} X(t, s) \sum_{i=0}^{i=M} B_i(s) u(v_i(s)) ds$$

$$+ \int_{t_0}^{t} X(t, s) \left(\int_{-h}^{0} A(\tau) x(s+\tau) d\tau \right) ds \quad \text{for } t \in [t_0, t_1] \tag{3.4}$$

$$x(t) = \phi(t) \quad \text{for } t \in (-\infty, t_0]$$

with initial state $z(t_0) = \{x(t_0); \phi, u_{t_0}\}$, and $x(t; t_0, \phi)$ is the solution of the uncontrolled linear differential equation

$$x'(t) = L(t, x_t).$$

In the sequel we shall also consider the linear control system with multiple lumped time varying delays in the in admissible controls and without delays in the state variables, described by the following differential state equation,

$$x'(t) = A(t) + \sum_{i=0}^{i=M} B_i(t) u(v_i(t)) \tag{3.5}$$

where $A(t)$ is $n \times n$ dimensional matrix whose elements are square integrable functions on $[t_0, \infty)$.

Using the time lead functions and the inequalities (3.3) we have

$$x(t_1) = x(t_1; t_0, \phi) + \sum_{i=0}^{i=m} \int_{v_i(t_0)}^{t_0} X(t, r_i(s)) B_i(r_i(s)) r'_i(s) u_{t_0}(s) ds$$

$$+ \sum_{i=m+1}^{i=M} \int_{v_i(t_0)}^{v_i(t_1)} X(t_1, r_i(s)) B_i(r_i(s)) r'_i(s) u_{t_0}(s) ds$$

$$+ \sum_{i=0}^{i=m} \int_{t_0}^{t_1} X(t_1, r_i(s)) B_i(r_i(s)) r'_i(s) u(s) ds \tag{3.6}$$

$$+ \int_{t_0}^{t_1} X(t_1, s) \left(\int_{-h}^{0} A(\tau) x(s+\tau) d\tau \right) ds$$

For brevity of considerations, let us introduce the following often used notations [31, 32].

$$x(t_1) = x(t_1; t_0, \phi) + \sum_{i=0}^{i=m} \int_{v_i(t_0)}^{t_0} X(t, r_i(s)) B_i(r_i(s)) r'_j(s) u_{t_0}(s) ds$$

$$+ \sum_{i=m+1}^{i=M} \int_{v_i(t_0)}^{v_i(t_1)} X(t_1, r_i(s)) B_i(r_i(s)) r'_j(s) u_{t_0}(s) ds$$

$$+ \sum_{i=0}^{i=m} \int_{t_0}^{t_1} X(t_1, r_i(s)) B_i(r_i(s)) r'_j(s) u(s) ds + \int_{t_0}^{t_1} X(t_1, s) \left(\int_{-h}^{0} A(\tau) x(s+\tau) d\tau \right) ds$$

(3.7)

Thus, $H(t, u_{t_0}) \in R^n$

$$x(t_1) = x(t_1; t_0, \phi) + \sum_{i=0}^{i=m} \int_{v_i(t_0)}^{t_0} X(t, r_i(s)) B_i(r_i(s)) r'_i(s) u_{t_0}(s) ds$$

$$+ \sum_{i=m+1}^{i=M} \int_{v_i(t_0)}^{v_i(t_1)} X(t_1, r_i(s)) B_i(r_i(s)) r'_i(s) u_{t_0}(s) ds$$

$$+ \sum_{i=0}^{i=m} \int_{t_0}^{t_1} X(t_1, r_i(s)) B_i(r_i(s)) r'_i(s) u(s) ds$$

$$+ \int_{t_0}^{t_1} X(t_1, s) \left(\int_{-h}^{0} A(\tau) x(s+\tau) d\tau \right) ds$$

(3.8)

Hence, using (3.7) we have

$$q(t_1, u_{t_0}) = x(t_1; t_0, \phi) + \sum_{i=0}^{i=m} \int_{v_i(t_0)}^{t_0} X(t, r_i(s)) B_i(r_i(s)) r'_i(s) u_{t_0}(s) ds$$

$$+ \sum_{i=m+1}^{i=M} \int_{v_i(t_0)}^{v_i(t_1)} X(t_1, r_i(s)) B_i(r_i(s)) r'_i(s) u_{t_0}(s) ds$$

$$+ \int_{t_0}^{t_1} X(t_1, s) \left(\int_{-h}^{0} A(\tau) x(s+\tau) d\tau \right) ds$$

(3.9)

3.2 System Description

Thus, $q(t_1, u_{t_0}) \in R^n$. Let us denote

$$G_m(t,s) = \sum_{i=0}^{i=m} X(t, r_i(s)) B_i(r_i(s)) r'_i(s) \qquad (3.10)$$

Thus, $G_m(t, s)$ is $n \times p$ dimensional matrix.

3.3 Controllability Conditions

Using the standard methods let us define the $n \times n$ dimensional relative controllability matrix at time t_1 for the linear dynamical control system (3.5) [31, 32].

$$W(t_0, t_1) = \int_{t_0}^{t_1} G_m(t_1, s) G_m^{tr}(t_1, s) ds \qquad (3.11)$$

where tr denotes matrix transpose.

Therefore, relative controllability matrix $W(t_0, t_1)$ is $n \times n$ dimensional symmetric matrix.

Theorem 3.1 *Linear delayed system* (3.1) *is relatively controllable in the time interval* $[t_0, t_1]$ *if and only if*

$$\text{rank } W(t_0, t_1) = n$$

Now let us consider simplified version of linear system (3, 5), i.e. system with multiple lumped time variable delays in the admissible controls, but without any delays in the state variables. Thus,

$$G_{m0}(t,s) = \sum_{i=0}^{i=m} X(t,s)) B_i(r_i(s)) r'_i(s)$$

Therefore, in this simplified case relative controllability matrix $W_0(t_0, t_1)$ is $n \times n$ dimensional symmetric matrix of the following form

$$W_0(t_0, t_1) = \int_{t_0}^{t_1} G_{m0}(t_1, s) G_{m0}^{tr}(t_1, s) ds$$

Hence, taking into account Theorem 3.1 we have the following necessary and sufficient condition for relative controllability given in the Corollary 3.1.

Corollary 3.1 *Linear delayed system* (3.5) *is relatively controllable in the time interval* $[t_0, t_1]$ *if and only if*

$$\text{rank } W_0(t_0, t_1) = n$$

Now, using controllability matrix minimum energy control problem can be easily solved.

3.4 Minimum Energy Control

Similarly, as for dynamical systems without delays, minimum energy control problem for delayed dynamical systems is strongly related to controllability problem. For controllable dynamical system (3.1) there exists generally many different controls which steer the system from a given initial state $x(t_0)$ to the final desired state x^1 at time $t_1 > 0$. Therefore, we may look for the control which is an optimal in the sense of the following performance index.

$$J(u) = \int_{t_0}^{t_1} \|u(t)\|_Q^2 dt$$

where

$$\|u(t)\|_Q^2 = u^{tr}(t) Q u(t)$$

and Q is an $m \times m$-dimensional constant symmetric and positive definite weighting matrix.

The performance index $J(u)$ defines the control energy in the time interval $[t_0, t_1]$ with the weight determined by the weighting matrix Q. The control u which minimizes the performance index $J(u)$ is called the minimum energy control. It should be mentioned, that the performance index $J(u)$ is a special case of the general quadratic performance index, and hence the existence of a minimizing control function is assured.

Now, let us consider linear time varying dynamical control systems, for which mathematical model is given by the following differential state equation with time varying parameter

$$x'(t) = A(t)x(t) + B(t)u(t) \qquad t \in [t_0, t_1] \qquad (3.11)$$

where similarly as before $x(t) \in R^n$ is a state vector, $u(t) \in R^m$ is an input vector, $A(t)$, and $B(t)$ are matrices of appropriate dimensions with continuous-time elements.

From [31] it is well known, that for a given initial state $x(t_0) \in R^n$ and control $u(t) \in R^m$, $t \geq t_0$, there exist unique solution $x(t; x(t_0), u) \in R^n$ of the state equation of the following form

3.4 Minimum Energy Control

$$x(t; x(t_0), u) = H(t, t_0)x(t_0) + \int_{t_0}^{t} H(t, s)B(s)u(s)ds$$

where $H(t, s)$ is fundamental matrix solution for equation.

Now, let us recall the most popular and frequently used fundamental definition of controllability for linear time-varying control systems with varying coefficients.

Definition 3.4 *Dynamical system* (3.1) *is said to be controllable in a given time interval* $[t_0, t_1]$ f *for every initial condition* $x(t_0)$ *and every vector* $x^1 \in R^n$, *given in finite time* t_1 *there exist control* $u(t) \in R^m$, $t \in [t_0, t_1]$, *such that* $x(t_1; x(t_1), u) = x^1$.

Similarly as for time invariant case this definition requires only that any initial state $x(t_0)$ can be steered to any final state x^1 at time t_1. The trajectory of the system is not specified. Furthermore, there is no constraints imposed on the control.

In order to formulate easily computable algebraic controllability criteria let us introduce the so called controllability matrix $W(t_0, t_1)$ defined as follows.

$$W(t_0, t_1) = \int_0^{t_1} H(t_0, t_1)B(t)B^{tr}(t)H(t_0, t_1)^{tr} dt,$$

Controllability matrix $W(t_0, t_1)$ is an $n \times n$-dimensional constant matrix and depends only on system parameters.

Theorem 3.3 *Dynamical system* (3.11) *is controllable if and only if*

$$\text{rank } W(t_0, t_1) = n$$

Corollary 3.2 *Dynamical system* (3.11) *is controllable if and only if the* $n \times n$-*dimensional symmetric matrix* $W(t_0, t_1)W(t_0, t_1)^{tr}$ *is nonsingular.*

Minimum energy control problem for time-varying system is defined similarly as for time-invariant case and of course is also strongly related to controllability problem.

Therefore, we may look for the control which steer the system from a given initial state $x(t_0)$ to the final desired state x^1 at time t_1 and is an optimal in the sense of the following performance index.

$$J(u) = \int_{t_0}^{t_1} \|u(t)\|_Q^2 dt$$

where

$$\|u(t)\|_Q^2 = u^{tr}(t)Qu(t)$$

and Q is an $m \times m$-dimensional constant symmetric and positive definite weighting matrix.

The performance index $J(u)$ defines the control energy in the time interval $[t_0, t_1]$ with the weight determined by the weighting matrix Q. The control u which minimizes the performance index $J(u)$ is called the minimum energy control. It should be mentioned, that the performance index $J(u)$ is a special case of the general quadratic performance index, and hence the existence of a minimizing control function is assured.

Therefore, the minimum energy control problem can be formulated as follows: for a given arbitrary initial state $x(t_0)$, arbitrary final state x^1, and finite time t_1, find an optimal control $u(t)$, $t \in [t_0, t_1]$, which transfers the state $x(t_0)$ to x^1 at time t_1 and minimizes the performance index $J(u)$.

In order to solve the minimum control problem and to present it in a readable compact form, let us introduce the following notation:

$$W_Q(t_0, t_1) = \int_{t_0}^{t_1} G_m(t_1, s) B(s) Q^{-1} B^{tr}(s) G_m^{tr}(t_1, s) ds$$

$W_Q(t_0, t_1)$ is constant $n \times n$-dimensional symmetric matrix

It should be pointed out, that since matrix Q is nonsingular, then matrix $W_Q(t_0, t_1)$ is nonsingular if and only if matrix $W(t_0, t_1)$ is nonsingular.

Let us define the admissible control of the following form

$$u^0(t) = Q^{-1} B(t)^{tr} [H(t_1, t)]^{tr} W_Q^{-1} [x^1 - H(t_1, t_0) x(t_0)] \quad (3.12)$$

Exact analytical solution of the minimum energy control problem for dynamical system (3.1) is given by the following Theorem.

Theorem 3.11 Let $u^1(t)$, $t \in [t_0, t_1]$ be any control that transfers initial state $x(0)$ to final state x^1 at time t_1, and let $u^0(t)$, $t \in [t_0, t_1]$ be the control defined by (3.10). Then the control $u^0(t)$ transfers the initial state $x(t_0)$ to a final state x^1 at time t_1 and

$$J(u^1) \geq J(u^0)$$

Moreover, the minimum value of the performance index corresponding to the optimal control u^0 is given by the following formula

$$J(u^0) = [x^1 - H(t_1, t) x(t_0)]^{tr} W_Q^{-1} [x^1 - H(t_1, t) x(t_0)]$$

It should be pointed out the Theorem 3.11 is proved under the following general assumptions:

1. dynamical system is linear,
2. there are no constraints in control,
3. there are no constraints posed on state variable $x(t)$,

3.4 Minimum Energy Control

4. dynamical system is controllable,
5. performance index $J(u)$ does not contain the state variable $x(t)$,
6. performance index is a quadratic with respect to control $u(t)$.

Moreover, it should be mentioned, that Theorem 3.11 can be proved by using only the fundamental properties of the norm and scalar product in the Hilbert spaces R^n and $L^2([0, t_1], R^m)$, without applying rather complicated results taken from the mathematical optimal control theory. This is possible only under the six general assumptions mentioned before.

The performance index $J(u)$ can also be used in determining the so called controllability measures, which characterize qualitatively the dynamical system (3.1). In the literature there are many different measures of controllability, generally depending on the eigenvalues of the matrix W_Q^{-1}.

Chapter 4
Controllability of Higher Order Linear Systems with Multiple Delays in Control

4.1 Introduction

In the present chapter finite-dimensional dynamical control systems described by linear higher-order ordinary differential state equations with multiple point delays in control are considered. Using algebraic methods, necessary and sufficient condition for relative controllability in a given time interval for linear dynamical system with multiple point delays in control is formulated and proved. This condition is generalization to relative controllability case some previous results concerning controllability of linear dynamical systems without multiple point delays in the control [21–23]. Proof of the main result is based on necessary and sufficient controllability condition presented in the paper [21] for linear systems without delays in control. Simple numerical example, which illustrates theoretical result is also given. Finally, some remarks and comments on the existing results for controllability of dynamical systems with delays in control are also presented.

4.2 System Description

In this chapter we study the linear higher-order control system with multiple point delays in the control described by the following ordinary differential state equation

$$x^{(N)}(t) = \sum_{i=0}^{i=N-1} A_i x^{(i)}(t) + \sum_{j=0}^{j=M} B_j u(t - h_j) \quad \text{for} \quad t \in [0, T], \quad (4.1)$$

with zero initial conditions:

$$x(0) = 0 \, u(t) = 0 \text{ for } \quad t \in [-h, 0) \quad (4.2)$$

where the state $x(t) \in R^n = X$ and the control $u(t) \in R^m = U$, A is $n \times n$ dimensional constant matrix, B_j, $j = 0, 1, 2, \ldots, M$. are $n \times m$. dimensional constant matrices,

$$0 = h_0 < h_1 < \ldots < h_j < \ldots < h_M = h$$

are constant delays.

The set of admissible controls for the dynamical control system (4.1) has the following form $U_{ad} = L_\infty([0, T])$.

Then for a given admissible control $u(t)$ there exists a unique solution $x(t; u)$ for $t \in [0, T]$, of the state Eq. (4.1) with zero initial condition (4.2). Moreover, the dynamical system (4.1) is equivalent to the first-order system

$$z^{(1)}(t) = Cz(t) + \sum_{j=0}^{j=M(T)} \tilde{D}_j u(t - h_j) \tag{4.3}$$

where

$$M(t) = j, \text{ for } h_j < t \leq h_{j+1}, j = 0, 1, 2, \ldots, (M-1)$$
$$M(t) = M, \text{ for } h_M < t$$

$$C = \begin{bmatrix} 0 & I & 0 & \cdots & 0 \\ 0 & 0 & I & \cdots & 0 \\ \vdots & \vdots & \vdots & \ddots & \vdots \\ 0 & 0 & 0 & \cdots & I \\ A_0 & A_1 & A_2 & \cdots & A_{N-1} \end{bmatrix}$$

$$z(t) = \begin{bmatrix} x(t) \\ x^{(1)}(t) \\ x^{(2)}(t) \\ \vdots \\ x^{(n-1)}(t) \end{bmatrix} \quad \tilde{B}_j = \begin{bmatrix} 0 \\ 0 \\ \vdots \\ 0 \\ B_j \end{bmatrix}$$

Hence the general solution of (4.1) is given by

$$x(t) = Z \exp(Ct) x(0) + Z \int_0^t \exp(C(t-\tau)) \sum_{j=0}^{j=M(t)} \tilde{B}_j u(t - h_j) d\tau$$

for $h_{M(T)} < t \leq h_{M(T)}$

where

$$Z = [I \mid 0 \mid 0 \mid \ldots \mid 0]$$

4.2 System Description

or equivalently

$$x(t) = Z \exp(Ct)x(0) + Z \int_0^t \exp(C(t-\tau))D_j(t)v(\tau)d\tau$$

where

$$D_j(t) = \begin{bmatrix} 0 & 0 & \cdots & 0 & 0 \\ 0 & 0 & \cdots & 0 & 0 \\ \cdots & \cdots & \ddots & 0 & 0 \\ 0 & 0 & \cdots & 0 & 0 \\ B_0 & B_1 & \cdots & B_{M(t)-1} & B_{M(t)} \end{bmatrix}$$

$$v(t) = \begin{bmatrix} u(t) \\ u(t-h_1) \\ \vdots \\ u(t-h_{M(t)-1}) \\ u(t-h_{M(t)}) \end{bmatrix}$$

In the next part of this section, we shall introduce certain notations and present some important facts from the general controllability theory of higher-order linear differential equations.

For the linear dynamical system with multiple point delays in the control (4.1), it is possible to define many different concepts of controllability. In the sequel we shall focus our attention on the relative controllability in the time interval $[0, T]$.

In order to do that, first of all let us introduce the notion of the attainable set at time $T > 0$ from zero initial conditions (4.2), denoted by $K_T(U_c)$ and defined as follows [22, 23].

$$K_T(U_c) = \{x \in X : x = x(T, u), u(t) \in U_c \text{ for a.e. } t \in [0, T]\} \quad (4.4)$$

where $x(t, u)$, $t > 0$ is the unique solution of the Eq. (4.1) with zero initial conditions (4.2) and a given admissible control u.

Now, using the concept of the attainable set given by the relation (4.4), let us recall the well known (see e.g. [22, 23] for details) definitions of relative controllability in $[0, T]$ for dynamical system (4.1).

Definition 4.1 Dynamical system (4.1) is said to be U_c-globally relative controllable in $[0, T]$ if $K_T(U_c) = R^n$.

Applying the Laplace transformation to the right-hand side of (4.1) yields the matrix polynomial

$$L(s) = Is^N - \sum_{i=0}^{i=N-1} A_i s^i$$

where s is a complex variable.

Moreover, let $\sigma(L)$ denotes the spectrum of the polynomial matrix $L(S)$, namely,

$$\sigma(L) = \{s \in C : \det L(s) = 0\}.$$

Let us consider polynomial matrix $L_T(s)$ as follows

$$L_T(s) = \left[L(s) \mid D(T)s^{nN-1} \mid D(T)s^{nN-2} \mid \ldots \mid D(T)s^2 \mid D(T)s \mid D(T) \right] \quad (4.5)$$

Let $p_1(s), p_2(s), \ldots, p_k(s), \ldots, p_K(s)$, where

$$K = \binom{n + mM(T)N}{n}$$

are determinants of all possible $n \times n$-dimensional submatrices of the matrix $L_T(s)$, which in fact are scalar polynomials of degree no more than nN. Therefore, for every $k = 1, 2, \ldots, K$ we have

$$p_k(s) = \left[1 \mid s \mid s^2 \mid \ldots \mid s^{nN} \right] r_k$$

where r_k is the vector of the corresponding coefficients of $p_k(s)$.

Finally, following [21] let us define Plücker matrix for the system (4.1) as follows

$$P(L(s), D(T)) = \left[r_1 \mid r_2 \mid \ldots \mid r_{K-1} \mid r_K \right]$$

4.3 Controllability Conditions

First of all, for completeness of considerations, let us recall necessary and sufficient conditions for relative controllability (see e.g. [21–23] for more details).

Lemma 4.1 *The following statements are equivalent.*

(i) System (4.1) is relatively controllable on $[0, T]$.
(ii) rank $[D(T), CD(T), C^2 D(T), \ldots, C^{nN-1} D(T)] = nN$
(iii) rank $[L(s), D(T)] = n$ for all $s \in \sigma(L)$

Now, let us formulate the main result of this chapter.

4.3 Controllability Conditions

Theorem 4.1 *The higher-order dynamical system* (4.1) *is relatively controllable on* $[0, T]$ *if and only if the Plücker matrix* $P(L(s), D(T))$ *has full row rank, i.e.*

$$rank\, P(L(s), D(T)) = nN + 1 \qquad (4.6)$$

Proof First of all, let us observe, that for dynamical systems with multiple delays in control (4.1) control we have piece-wise constant matrix $D(t)$ instead constant matrix B as for dynamical systems without delays in control. Therefore, following [1], the $(nN + 1) \times K$ dimensional matrix $P(L(s), D(T))$ essentially depends on T, and for $h_{M(T)} < t \le h_{M(T)}$ has the form (4.5).

Thus, using necessary and sufficient condition for controllability of systems without delays in control [31] we obtain equality (4.6).

Corollary 4.1 *The second-order dynamical system* (4.1) *is relatively controllable on* $[0, T]$ *if and only if the Plücker matrix* $P(L(s), D(T))$ *has full row rank, i.e.,*

$$rank\, P(L(s), D(T)) = 2n + 1 \qquad (4.7)$$

Example 4.1 Let us consider the following simple illustrative example. Let the second-order linear finite-dimensional dynamical control system defined on a given time interval $[0, T]$, has the following form

$$\begin{aligned} x_1^{(2)}(t) &= -x_2^{(1)}(t) - x_1(t) + x_2(t) + u(t) \\ x_2^{(2)}(t) &= -x_1^{(1)}(t) + x_1(t) + x_2(t) + u(t - h) \end{aligned} \qquad (4.8)$$

Therefore, $n = 2$, $m = 1$, $M = 1$, $0 < h$, $x(t) = (x_1(t), x_2(t))^{tr} \in R^2$ $U = R$, and using the notations given in the previous sections matrices A_0, A_1 and B_0, B_1 have the following form

$$A_0 = \begin{bmatrix} -1 & 1 \\ +1 & +1 \end{bmatrix}$$

$$A_1 = \begin{bmatrix} 0 & -1 \\ -1 & 0 \end{bmatrix}$$

$$B_0 = \begin{bmatrix} 1 \\ 0 \end{bmatrix}$$

$$B_1 = \begin{bmatrix} 0 \\ 1 \end{bmatrix}$$

Therefore, matrix $L_T(s)$ has the following form

$$L(s) = Is^2 - A_1 s - A_0 = \begin{bmatrix} s^2 + 1 & s - 1 \\ s - 1 & s^2 - 1 \end{bmatrix}$$

Hence,

$$L_T(s) = [L(s) \vdots D(T)s \vdots D(T)] = [L(s) \vdots B_0 s \vdots B_0]$$
$$= \begin{bmatrix} s^2+1 & s-1 & s & 1 \\ s-1 & s^2-1 & 0 & 0 \end{bmatrix}$$
$$\text{for } 0 < T < h$$

In this case we have $n = 2$, $m = 1$, $M(T) = 1$, $N = 2$. Hence $K = 6$.

Thus, we have 6 determinants of 2×2-dimensional submatrices of the following form:

$$p_1(s) = s^4 - s^2 + 2s - 2 = \begin{bmatrix} 1 & s & s^2 & s^3 & s^4 \end{bmatrix} \begin{bmatrix} -2 \\ 2 \\ -1 \\ 0 \\ 1 \end{bmatrix}$$
$$= \begin{bmatrix} 1 & s & s^2 & s^3 & s^4 \end{bmatrix} r_1$$

and hence

$$r_1 = [-2, 2, -1, 0, 1]^{tr}$$

Similarly, without going into details it can be shown that:

$$p_2(s) = -s^2 + 1, \quad r_2 = [1, 0, -1, 0, 0]^{tr}$$
$$p_3(s) = -s + 1, \quad r_3 = [1, -1, 0, 0, 0]^{tr}$$
$$p_4(s) = -s^3 + s, \quad r_4 = [0, 1, 0, -1, 0]^{tr}$$
$$p_5(s) = -s^2 + 1, \quad r_5 = [1, 0, -1, 0, 0]^{tr}$$
$$p_6(s) = 0, \quad r_6 = [0, 0, 0, 0, 0]^{tr}$$

Therefore,

$$\text{rank}P(L(S), B_0) = \text{rank} \begin{bmatrix} -2 & 1 & 1 & 0 & 1 & 0 \\ 2 & 0 & -1 & 1 & 0 & 0 \\ -1 & -1 & 0 & 0 & -1 & 0 \\ 0 & 0 & 0 & -1 & 0 & 0 \\ 1 & 0 & 0 & 0 & 0 & 0 \end{bmatrix} = 4 < 5 = nN+1$$

Therefore system is not relatively controllable on $[0, T]$ for $T < h$.
On the other hand $s = 1$ is an eigenvalue of the matrix $L(s)$.
Moreover, we have

4.3 Controllability Conditions

$$rank\begin{bmatrix}L(s) & \vdots & B_0\end{bmatrix} = rank\begin{bmatrix} s^2+1 & s-1 & 1 \\ s-1 & s^2-1 & 0 \end{bmatrix}$$

$$= rank\begin{bmatrix}L(1) & \vdots & B_0\end{bmatrix} = rank\begin{bmatrix} 2 & 0 & 1 \\ 0 & 0 & 0 \end{bmatrix} = 1 < 2$$

and system is not relatively controllable on $[0, T]$ for $T < h$.
However, for $T > h$ we have

$$L_T(s) = \begin{bmatrix}L(s) & \vdots & D(T)s & \vdots & D(T)\end{bmatrix} = \begin{bmatrix}L(s) & \vdots & B_0 s & \vdots & B_0\end{bmatrix}$$
$$= \begin{bmatrix} s^2+1 & s-1 & s & 1 \\ s-1 & s^2-1 & 0 & 0 \end{bmatrix}$$

In this case we have $n = 2$, $m = 1$, $M(T) = 2$, $N = 2$. Hence, $K = 15$. Thus, we have 15 determinants of 2×2 dimensional submatrices.

Computing determinant $p_k(s)$ and coefficients vectors r_k we form 5×15-dimensional matrix $P(L(s), B_1, B_0)$, for which

$$rank P(L(S), B_1, B_0) = 5 = nN + 1$$

Hence, system is relatively controllable on $[0, T]$ for $h < T$.
On the other hand $s = 1$ is an eigenvalue of the matrix $L(s)$.
Moreover, we have

$$rank\begin{bmatrix}L(s) & \vdots & B_1 & \vdots & B_0\end{bmatrix} = rank\begin{bmatrix} s^2+1 & s-1 & 0 & 1 \\ s-1 & s^2-1 & 1 & 0 \end{bmatrix}$$

$$= rank\begin{bmatrix}L(1) & B_1 & B_0\end{bmatrix} = rank\begin{bmatrix} 2 & 0 & 0 & 1 \\ 0 & 0 & 1 & 0 \end{bmatrix} = 2$$

and system is relatively controllable on $[0, T]$ for $h < T$.

In this Chapter necessary and sufficient conditions for relative controllability on o given time interval for higher-order finite-dimensional dynamical control systems with multiple point delays in the control have been formulated and proved. In the proof of the main result previous has been used. The important feature of the new rank condition is that it does not require computation of eigenvalues of the considered dynamical system.

These conditions extend to the case of relative controllability and dynamical control systems with delays in control the results published in [31] for linear systems without delays in control.

Chapter 5
Constrained Controllability of Semilinear Systems with Multiple Constant Delays in Control

5.1 Introduction

In the present chapter, we shall consider constrained local relative controllability problems for finite-dimensional semilinear dynamical systems with multiple point delays in the control described by ordinary differential state equations.

Let us recall, that semilinear dynamical control systems contain linear and pure nonlinear parts in the differential state equations [63, 80]. More precisely, we shall formulate and prove sufficient conditions for constrained local relative controllability in a prescribed time interval for semilinear dynamical systems with multiple point delays in the control which nonlinear term is continuously differentiable near the origin.

It is generally assumed that the values of admissible controls are in a given convex and closed cone with vertex at zero, or in a cone with nonempty interior. Proof of the main result is based on a so called generalized open mapping theorem presented in the paper [68].

Moreover, necessary and sufficient conditions for constrained global relative controllability of an associated linear dynamical system with multiple point delays in control are discussed.

Simple numerical example which illustrates theoretical considerations is also given. Finally, some remarks and comments on the existing results for controllability of nonlinear dynamical systems are also presented.

Roughly speaking, it will be proved that under suitable assumptions constrained global relative controllability of a linear associated approximated dynamical system implies constrained local relative controllability near the origin of the original semilinear dynamical system.

This is a generalization to constrained controllability case some previous results concerning controllability of linear dynamical systems with multiple point delays in the control and with unconstrained controls [31–33].

5.2 System Description

In this paper we study the semilinear control system with multiple point delays in the control described by the following ordinary differential state equation

$$\dot{x}(t) = Ax(t) + F(x(t)) + \sum_{j=0}^{j=M} B_j u(t - h_j) \quad \text{for} \quad t \in [0, T], T > h \quad (5.1)$$

with zero initial conditions:

$$x(0) = 0 \quad u(t) = 0 \quad \text{for} \quad t \in [-h, 0) \quad (5.2)$$

where the state $x(t) \in R^n = X$
the control $u(t) \in R^m = U$,
A is $n \times n$-dimensional constant matrix,
B_j, $j = 0, 1, 2, \ldots, M$. are $n \times m$.- dimensional constant matrices,

$$0 = h_0 < h_1 < \ldots < h_j < \ldots < h_M = h$$

are constant delays.

Moreover, let us assume that the nonlinear mapping $F: X \to X$ is continuously differentiable near the origin and such that $F(0) = 0$.

In practice admissible controls are always required to satisfy certain additional constraints. Generally, for arbitrary control constraints it is rather very difficult to give easily computable criteria for constrained controllability. However, for some special cases of the constraints it is possible to formulate and prove simple algebraic constrained controllability conditions.

Therefore, we assume that the set of values of controls $U_c \subset U$ is a given closed and convex cone with nonempty interior and vertex at zero. Then the set of admissible controls for the dynamical control system (5.1) has the following form $U_{ad} = L_\infty([0, T], U_c)$.

Then for a given admissible control u(t) there exists a unique solution $x(t; u)$ for $t \in [0, T]$, of the state Eq. (5.1) with zero initial condition (5.2) described by the integral formula [63, 80].

$$x(t; u) = \int_0^t S(t - s)(F(x(s; u)) + \sum_{j=0}^{j=M} B_j u(t - h_j) ds \quad (5.3)$$

where $S(t) = exp(At)$ is $n \times n$-dimensional transition matrix for the linear part of the semilinear control system (5.1).

For the semilinear dynamical system with multiple point delays in the control (6.1), it is possible to define many different concepts of controllability. In the sequel we shall focus our attention on the so called constrained relative controllability in the time interval [0, T].

5.2 System Description

In order to do that, first of all let us introduce the notion of the attainable set at time $T > 0$ from zero initial conditions (5.2), denoted by $K_T(U_c)$ and defined as follows [31, 32, 69].

$$K_T(U_c) = \{x \in X : x = x(T, u), u(t) \in U_c \quad for \quad a.e. \quad t \in [0, T]\} \quad (5.4)$$

where $x(t, u)$, $t > 0$ is the unique solution of the Eq. (5.1) with zero initial conditions (5.2) and a given admissible control $u(t)$, $t > 0$. Let us observe, that under the assumptions stated on the nonlinear term F such solution always exists [63, 80].

Now, using the concept of the attainable set given by the relation (5.4), let us recall the well known (see e.g. [31–33] definitions of constrained relative controllability in [0, T] for dynamical system (5.1).

Definition 5.1 Dynamical system (5.1) is said to be U_c-locally relative controllable in [0, T] if the attainable set $K_T(U_c)$ contains a certain neighborhood of zero in the space X.

Definition 5.2 Dynamical system (5.1) is said to be U_c-globally relative controllable in [0, T] if $K_T(U_c) = X$.

Finally, it should be stressed, that in a quite similar way, we may define constrained global and local absolute controllability [3]. However, since in this case the state space is in fact infinite-dimensional, then it is necessary to distinguish between exact and approximate absolute controllability.

5.3 Preliminaries

In this section, we shall introduce certain notations and present some important facts from the general theory of nonlinear operators.

Let U and X be given Banach spaces and $g(u):U \to X$ be a mapping continuously differentiable near the origin 0 of U.

Let us suppose for convenience that $g(0) = 0$. It is well known from the implicit-function theorem (see e.g. [68]) that, if the derivative $Dg(0):U \to X$ maps the space U onto the whole space X, then the nonlinear map g transforms a neighborhood of zero in the space U onto some neighborhood of zero in the space X.

Now, let us consider the more general case when the domain of the nonlinear operator g is Ω, an open subset of U containing 0. Let U_c denote a closed and convex cone in U with vertex at 0.

In the sequel, we shall use for controllability investigations some property of the nonlinear mapping g which is a consequence of a generalized open-mapping theorem [68]. This result seems to be widely known, but for the sake of completeness we shall present it here, though without proof and in a slightly less general form sufficient for our purpose.

Lemma 5.1 [68] *Let X, U, U_c, and Ω be as described above. Let $g:\Omega \to X$ be a nonlinear mapping and suppose that on Ω nonlinear mapping g has derivative Dg, which is continuous at 0. Moreover, suppose that g(0)=0 and assume that linear map Dg(0) maps U_c onto the whole space X.*

Then there exist neighborhoods $N_0 \subset X$ about $0 \in X$ and $M_0 \subset \Omega$ about $0 \in U$ such that the nonlinear equation $x = g(u)$ has, for each $x \in N_0$, has at least one solution $u \in M_0 \cap U_c$, where $M_0 \cap U_c$ is a so called conical neighborhood of zero in the space U.

Lemma 5.2 *Let $D_u x$ denotes derivative of x with respect to u. Moreover, if x(t; u) is continuously differentiable with respect to its u argument, we have for $v \in L_\infty([0, T], U)$*

$$D_u x(t; u)(v) = z(t, u, v)$$

where the mapping $t \to z(t, u, v)$ is the solution of the linear ordinary equation

$$\dot{z}(t) = Az(t) + D_x(F(x; u))z(t) + \sum_{j=0}^{j=M} B_j v(t - h_j) \qquad (5.5)$$

with zero initial conditions $z(0, u, v) = 0$ and $v(t) = 0$ for $t \in [-h, 0)$.

Proof Using formula (5.3) and the well known differentiability results we have

$$D_u x(t; u) = \int_0^t D_u(S(t-s)(F(x(t;u)) + \sum_{j=0}^{j=M} B_j u(t-h_j))ds$$

$$= \int_0^t S(t-s)D_u(F(x(t;u)) + \sum_{j=0}^{j=M} B_j u(t-h_j))ds$$

$$= \int_0^t S(t-s)D_x F(x(t;u))D_u x(t;u)ds + \int_0^t S(t-s)(\sum_{j=0}^{j=M} B_j u(t-h_j))ds$$

$$(5.6)$$

Differentiating equality (5.6) with respect to the variable *t*, we have

$$(d/dt)D_u x(t; u)v = D_x F(x(t; u))D_u x(t; u)v + \sum_{j=0}^{j=M} B_j u(t - h_J)v$$

$$+ \int_0^t (d/dt)S(t-s) \sum_{j=0}^{j=M} B_j u(t-h_J))dsv \qquad (5.7)$$

$$+ \int_0^t (d/dt)S(t-s)D_x F(x(s;u))D_u x(s;u)dsv$$

5.3 Preliminaries

Therefore, since by assumption $S(t)$ is a differentiable semigroup then

$$(d/dt)S(t-s) = AS(t-s)$$

and we have

$$\dot{z}(t) = D_x F(x(t;u))z(t) + \left(\int_0^t AS(t-s)\sum_{j=0}^{j=M} B_j v(t-h_j))ds\right)$$
$$+ \left(\int_0^t AS(t-s)D_x F(x(s;u))z(s)ds\right) \tag{5.8}$$

On the other hand solution of the Eq. (5.1) has the following integral form

$$z(t) = \int_0^t S(t-s)(D_x F(x(s;u))z(s) + \sum_{j=0}^{j=M} B_j v(t-h_j))ds \tag{5.9}$$

Therefore, differential Eq. (5.4) can be expressed as follows

$$\dot{z}(t) = Az(t) + D_x F(x(t;u))z(t) + \sum_{j=0}^{j=M} B_j v(t-h_j)$$

Hence, Lemma 5.2 follows.

5.4 Controllability Conditions

In this Section we shall study constrained local relative controllability near the origin in the time interval $[0, T]$ for semilinear dynamical system (5.1) using the associated linear dynamical system with multiple point delays in the control

$$\dot{z}(t) = Cx(t) + \sum_{j=0}^{j=M} B_j u(t-h_J) \quad \text{for} \quad t \in [0, T] \tag{5.10}$$

with zero initial condition $z(0) = 0$, $u(t) = 0$, for $t \in [-h, 0)$, where

$$C = A + D_x F(0)$$

The main result is the following sufficient condition for constrained local relative controllability of the semilinear dynamical system (5.1).

Theorem 5.1 *Suppose that*

(i) $F(0) = 0$,
(ii) $U_c \subset U$ *is a closed and convex cone with vertex at zero*,
(iii) *The associated linear control system with multiple point delays in the control (5.1) is U_c-globally relative controllable in $[0, T]$.*

Then, the semilinear dynamical control system with multiple point delays in the control (5.1) is U_c-locally relative controllable in $[0, T]$.

Proof Let us define for the nonlinear dynamical system (5.1) a nonlinear map

$$g : L_\infty([0, T], U_c) \to X \quad \text{by} \quad g(u) = x(T, u).$$

Similarly, for the associated linear dynamical system (5.1), we define a linear map

$$H : L_\infty([0, T], U_c) \to X \text{ by } Hv = z(T, v).$$

By the assumption (iii) the linear dynamical system (5.5) is U_c-globally relative controllable in $[0, T]$.

Therefore, by the Definition 5.2 the linear operator H is surjective, i.e., it maps the cone U_{ad} onto the whole space X. Furthermore, by Lemma 5.2 we have that $Dg(0) = H$.

Since U_c is a closed and convex cone, then the set of admissible controls $U_{ad} = L_\infty([0, T], U_c)$ is also a closed and convex cone in the function space $L_\infty([0, T], U)$. Therefore, the nonlinear map g satisfies all the assumptions of the generalized open mapping theorem stated in the Lemma 5.1.

Hence, the nonlinear map g transforms a conical neighborhood of zero in the set of admissible controls U_{ad} onto some neighborhood of zero in the state space X. This is by Definition 5.1 equivalent to the U_c-local relative controllability in $[0, T]$ of the semilinear dynamical control system (5.1). Hence, our theorem follows.

Let us observe, that in practical applications of the Theorem 5.1, the most difficult problem is to verify the assumption (iii) about constrained global controllability in a given time interval of the linear dynamical system with multiple point delays in the control (5.1), [31–33, 66, 72]. In order to avoid this disadvantage, we may use the Theorems and Corollaries given in the next section.

5.5 Constrained Controllability Conditions for Linear Systems

The main result is the following necessary and sufficient condition for constrained relative controllability of the linear dynamical system (5.1).

Theorem 5.2 *Linear control system with multiple point delays in the control (5.5) is U_c-relative controllable in $[0, T]$ for*

5.5 Constrained Controllability Conditions for Linear Systems

$$h_k < T \leq h_{k+1}, \, k = 0, 1, 2, \ldots, M, h_{M+1} = +\infty,$$

if and only if the linear dynamical control system without point delays in control.

$$\dot{x}(t) = Cx(t) + \tilde{B}_k v(t) \tag{5.11}$$

where

$$\tilde{B}_k = \begin{bmatrix} B_0 & | & B_1 & | & \ldots & | & B_j & | & \ldots & | & B_k \end{bmatrix}$$

is V_c-controllable in time interval $[0, T]$, where $v(t) \in V_{ad} = L_\infty([0, T], V_c)$, and $V_c = U_c \times U_c \times \ldots \times U_c \in R^{m(k+1)}$ is a given closed and convex cone with nonempty interior and vertex at zero.

Proof Let First of all, taking into account zero initial conditions (5.6) and the form of time-variable delays, changing the order of integration, let us transform equality (5.7) as follows

$$x(t; u) = \int_0^t \exp(C(t-s)) \left(\sum_{j=0}^{j=M} B_j u(s - h_j) \right) ds$$

$$= \sum_{j=0}^{j=M} \int_0^t \exp(C((t-s)) B_j u(s - h_j)) ds$$

$$= \sum_{j=0}^{j=k} \int_0^{t-h_j(t)} \exp(C((t-s+h)) B_j u(s)) ds$$

for t satisfying inequalities

$$h_j < t \leq h_{j+1}, \, j = 0, 1, 2, \ldots, k-1$$

Moreover, for simplicity of notation, let us denote:

$$\tilde{B}_k = \begin{bmatrix} B_0 & | & B_1 & | & \ldots & | & B_j & | & \ldots & | & B_k \end{bmatrix}$$

where $\tilde{B}_k$ are $n \times m(k+1)$-constant dimensional matrices for

$$h_k < t \leq h_{k+1}, \, k = 0, 1, 2, \ldots, M, \text{ where } h_{M+1} = +\infty$$

Let us observe that the matrices $exp(C(t-s + h_j(s)))$ are always nonsingular, and moreover, therefore, they do not change controllability property, and relative controllability linear system with delays in control (5.1) is in fact equivalent to controllability of the following linear system without delays in the control (5.1) [32].

$$\dot{x}(t) = Cx(t) + \tilde{B}_k v(t) \quad \text{for } h_k < t \leq h_{k+} \quad k = 0, 1, 2, \ldots, \tag{5.12}$$

where $v(t) \in V_{ad} = L_\infty([0, T], V_c)$, and $V_c = U_c \times U_c \times \ldots \times U_c \in R^{m(k+1)}$ is a given closed and convex cone with nonempty interior and vertex at zero.

Hence, Theorem 5.2 follows.

Now, using results concerning constrained controllability of linear system without delays (5.6) (see e.g. [31–33] for more details) we shall formulate and prove necessary and sufficient conditions for constrained relative controllability of linear dynamical systems with delays in control (5.1).

Remark 5.1 The monograph [32] contains quite general models of time-varying linear finite-dimensional dynamical systems both with distributed and time-varying multiple point delays in the control.

Theorem 5.3 *Suppose the set U_c is a cone with vertex at zero and a nonempty interior in the space R^m. Then the linear dynamical control system (5.5) is U_c-relatively controllable in $[0, T]$, $h_k < T \leq h_{k+1}$, $k = 0, 1, 2,\ldots, M$ if and only if*

(i) it is relative controllable in $[0, T]$ without any constraints, i.e.,

$$\text{rank}\left[B_0, B_1, \ldots, B_k, CB_0, CB_1, \ldots, CB_k, C^2 B_0, C^2 B_1, \ldots, C^2 B_k, \ldots, C^{n-1} B_0, C^{n-1} B_1, \ldots, C^{n-1} B_k\right] = n,$$

(ii) there is no real eigenvector $w \in R^n$ of the matrix C^{tr} satisfying

$$w^{tr} \tilde{B}_k v \leq 0 \quad \text{for all } v \in V_c.$$

Proof First of all let us recall, that by Theorem 5.2 constrained relative controllability of linear systems with delays in control given by linear state Eq. (5.5) is equivalent to constrained controllability of linear system without delays (5.5). On the other site, it is well known [31, 32] that necessary and sufficient conditions for constrained controllability of system (5.5) are exactly conditions (i) and (ii) given above. Hence our Theorem 5.3 follows.

Let us observe, that for a special case when $T < h_1$, controllability problem in $[0, T]$ may be reduced to the dynamical system without delays in the control [33].

$$\dot{x}(t) = Cx(t) + B_0 u(t) \tag{5.13}$$

Therefore, we can formulate the following Corollary.

Corollary 5.1 [31–33]. *Suppose that $T < h_1$ and the assumptions of Theorem 5.1 are satisfied. Then the linear dynamical control system (5.1) is U_c-controllable in $[0, T]$ if and only if it is controllable without any constraints, i.e.,*

5.5 Constrained Controllability Conditions for Linear Systems

$$rank\left[B_0, CB_0, C^2B_0, \ldots, C^{n-1}B_0\right] = n,$$

and there is no real eigenvector $w \in R^n$ of the matrix C^{tr} satisfying $w^{tr}B_0 u \leq 0$ for all $u \in U_c$.

Moreover, for the linear dynamical control system (5.1), with matrix A having only complex eigenvalues Theorem 5.3 reduces to the simple following Corollary.

Corollary 5.2 [31–33]. *Suppose that matrix C has only complex eigenvalues. Then the linear dynamical control system (5.1) is U_c-relative controllable in $[0, T]$, $h_k < T \leq h_{k+1}$, $k = 0, 1, 2, \ldots, M$ if and only if it is relative controllable without any constraints.*

It should be pointed out, that for scalar admissible controls, i.e. for $m = 1$, and $U_c = R^+$, the condition (ii) of Theorem 5.3 may holds when the matrix A has only complex eigenvalues. Hence we have the following Corollary.

Corollary 5.3 [1, 2, 5]. *Suppose that $m = 1$. Then the linear dynamical control system (5.1) is U_c-relative controllable in $[0, T]$, $h(T) < T$, if and only if it is relative controllable in $[0, T]$ without any constraints, and matrix A has only complex eigenvalues.*

Finally, let us consider the simplest case, when the final time is small enough, i.e. $T \leq h_1(T)$, and the matrix A has only complex eigenvalues. In this case relative constrained controllability problem in $[0, T]$ reduces to a very well known in the literature matrix rank condition [31–33].

Corollary 5.4 [5, 31, 32, 35]. *Suppose that $T \leq h_1(T)$, $U_c = R^+$ and matrix A has only complex eigenvalues. Then the linear dynamical control system (5.1) is U_c-controllable in $[0, T]$ if and only if it is controllable without any constraints, i.e.*

$$rank\left[B_0, AB_0, A^2B_0, \ldots, A^{n-1}B_0\right] = n,$$

Remark 5.2 From the above theorems and corollaries directly follows, that constrained relative controllability strongly depends on the delays and on the length of the time interval $[0, T]$.

Example 5.1 Let us consider the following simple illustrative example. Let the semilinear finite-dimensional dynamical control system defined on a given time interval $[0, T]$, $T > h = h_2$, has the following form

$$\begin{aligned} \dot{x}_1(t) &= -x_2(t) + u(t - h_1) \\ x_2(t) &= \sin x_1(t) + u(t - h_2) \end{aligned} \quad (5.14)$$

Therefore, $n = 2$, $m = 1$, $M = 2$, $0 = h_0 < h_1 < h_2 = h$, $x(t) = (x_1(t), x_2(t))^{tr} \in R^2 = X$, $U = R$, and using the notations given in the previous sections matrices A and B and the nonlinear mapping F have the following form

$$A = \begin{bmatrix} 0 & -1 \\ 0 & 0 \end{bmatrix}$$

$$B_0 = \begin{bmatrix} 0 \\ 0 \end{bmatrix} \quad B_1 = \begin{bmatrix} 1 \\ 0 \end{bmatrix} \quad B_2 = \begin{bmatrix} 0 \\ 1 \end{bmatrix}$$

$$F(x) = F(x_1, x_2) = \begin{bmatrix} 0 \\ \sin x_1 \end{bmatrix}$$

Moreover, let the cone of values of controls $U_c = R^+$, and the set of admissible controls $U_{ad} = L_\infty([0, T], R^+)$. Hence, we have

$$F(0) = F(0, 0) = \begin{bmatrix} 0 \\ 0 \end{bmatrix}$$

$$D_x F(x) = \begin{bmatrix} 0 & 0 \\ \cos x_1 & 0 \end{bmatrix}$$

$$D_x F(0) = \begin{bmatrix} 0 & 0 \\ 1 & 0 \end{bmatrix}$$

$$C = A + D_x F(0) = \begin{bmatrix} 0 & -1 \\ 1 & 0 \end{bmatrix}$$

Therefore, the matrix C has only complex eigenvalues and

$$rank[B_0, B_1, B_2, CB_0, CB_1, CB_2] = rank \begin{bmatrix} 0 & 1 & 0 & 0 & 0 & -1 \\ 0 & 0 & 1 & 0 & 0 & 0 \end{bmatrix}$$
$$= 2 = n$$

Hence, both assumptions of the Corollary 5.2 are satisfied and therefore, the associated linear dynamical control system (5.1) is R^+-globally controllable in a given time interval $[0, T]$. Then, all the assumptions stated in the Theorem 5.1 are also satisfied and thus the semilinear dynamical control systems (5.1) is R^+-locally controllable in $[0, T]$. However, it should be mentioned, that since

$$rank[B_0, B_1, CB_0, CB_1] = rank \begin{bmatrix} 0 & 1 & 0 & 0 \\ 0 & 0 & 0 & 0 \end{bmatrix} = 1 < n = 2$$

Then the semilinear dynamical control system (5.1) may be not relative controllable in the interval $[0, T]$, for $T < h_2$, even for unconstrained controls.

In this chapter sufficient conditions for constrained local relative controllability near the origin for semilinear finite-dimensional dynamical control systems with multiple point delays in the control have been formulated and proved. In the proof of the main result generalized open mapping theorem [68] has been used.

These conditions extend to the case of constrained relative controllability of finite-dimensional semilinear dynamical control systems the results published in [31–33] for unconstrained nonlinear systems.

The method presented in the paper is quite general and covers wide class of semilinear dynamical control systems. Therefore, similar constrained controllability results may be derived for more general class of semilinear dynamical control systems. For example, it is possible to extend sufficient constrained controllability conditions given in the previous section for semilinear dynamical control systems with distributed delay in the control or with point delays in the state variables and for the discrete-time semilinear control systems.

Finally, is should be mentioned, that other different controllability problems both for linear and nonlinear dynamical control systems have been also considered in the papers [1, 2, 9–11].

5.6 Positive Controllability of Positive Dynamical Systems

The present chapter is devoted to a study of constrained controllability and controllability for linear dynamical systems if the controls are taken to be nonnegative. In analogy to the usual definition of controllability it is possible to introduce the concept of positive controllability.

We shall concentrate on approximate positive controllability for linear infinite-dimensional dynamical systems when the values of controls are taken from a positive closed convex cone and the operator of the system is normal and has pure discrete point spectrum. The special attention is paid for positive infinite-dimensional linear dynamical systems. General approximate constrained controllability results are then applied for distributed parameter dynamical systems described by linear partial differential equations of parabolic type with different kinds of boundary conditions. Several remarks and comments on the relationships between different concepts of controllability are given. Finally, simple numerical illustrative example is also presented.

In the next part of this chapter we shall introduce some basic notations and definitions which will be used in the parts of the paper.

Throughout the nest part of chapter we use X to denote infinite dimensional separable real Hilbert space. By $L^p([0, t], R^m)$, $1 \leq p \leq \infty$ we denote the space of all p-integrable functions on $[0, t]$ with values in R^m, and $L^p_\infty([0, \infty), R^m)$ the space of all locally p-integrable functions on $[0, \infty)$ with values in R^m.

We define an order $\leq$ in the space X such that $(X, \leq)$ is a lattice and the ordering is compatible with the structure of X, i.e. X is an ordered vector space. This imply that the set $X^+ = \{x \in X: x \geq 0\}$ is a convex positive cone with vertex at zero. Moreover, it follows that $x_1 \leq x_2$ if and only if $x_2 - x_1 \in X^+$. An element $x \in X^+$ is called positive, and we write $x > 0$ if x is positive and different from zero. Moreover, an element $x^* \in X^+$ is called strictly positive, and we write $x^* \gg 0$ if $\langle x^*, x \rangle_X > 0$ for all $x > 0$. An ordered vector space X is called a vector lattice if any two elements x_1, x_2 in X have a supremum and an infimum denoted by $\sup\{x_1, x_2\}$, respectively, $\inf\{x_1, x_2\}$. For an element x of vector lattice we write $|x| = \sup\{x, -x\}$ and call it the absolute value of x. We call two elements x_1, x_2 of vector lattice X orthogonal, if $\inf\{|x_1|, |x_2|\} = 0$. Linear form $w \in X$ is called positive ($w \geq 0$) if $\langle w, x \rangle_X \geq 0$ for all $x \geq 0$ and strictly positive ($w \gg 0$) if $\langle w, x \rangle_X > 0$ for all $x > 0$. Relevant examples of vector lattices with a strictly positive linear form are given by the following spaces of practical interest: R^n and $L^2(\Omega, R)$, where Ω is a measurable subset of R^n.

Linear bounded operator F from a vector lattice X into a vector lattice V is called positive, i.e. $F \geq 0$, if $Fx \geq 0$ for $x \geq 0$. Therefore, positive operator F maps positive cone X^+ into positive cone V^+. Let $S(t): X \to X$, $t \geq 0$, be a strongly continuous semigroup of bounded linear operators. We call the semigroup positive, i.e. $S \geq 0$, if X is a vector lattice and $S(t)$ is a positive operator for every $t \geq 0$.

If set $M \subseteq X$, we define the polar cone by $M^o = \{w \in X, \langle w, x \rangle_X \leq 0$ for all $x \in M\}$. The closure, the convex hull and the interior are denoted respectively, by cl M, co M and int M.

Let us consider linear infinite-dimensional time-invariant control system of the following form

$$x'(t) = Ax(t) + Bu(t) \tag{5.15}$$

here $x(t) \in X$—infinite-dimensional separable Hilbert space which is a vector lattice with a strictly positive linear form.

B is a linear bounded operator from the space R^m into X. Therefore operator

$$B = [b_1, b_2, \ldots, b_j, \ldots, b_m] \text{ and}$$

$$Bu(t) = \sum_{j=1}^{j=m} b_j u_j(t)$$

where

$$b_j \in X \text{ for } j = 1, 2, \ldots, m, \text{ and}$$
$$u(t) = [u_1(t), u_2(t), \ldots, u_j(t), \ldots, u_m(t)]^{tr}.$$

We would like to emphasize that the assumption that linear operator B is bounded, rules out the application of our theory to boundary control problems, because in this situation B is typically unbounded.

$A: X \supset D(A) \to X$ is normal generally unbounded linear operator with compact resolvent $R(s, A)$ for all s, in the resolvent set $\rho(A)$. Then operator A has the following properties [30, 31]:

(1) Operator A has only pure discrete point spectrum $\sigma_p(A)$ consisting entirely with isolated eigenvalues s_i, $i = 1, 2, 3, \ldots$ Moreover, each eigenvalue s_i has finite multiplicity $n_i < \infty$, $i = 1, 2, 3, \ldots$ equal to the dimensionality of the corresponding eigenmanifold.

5.6 Positive Controllability of Positive Dynamical Systems

(2) The eigenvectors $x_{ik} \in D(A)$, $i = 1, 2, 3,\ldots$ $k = 1, 2, 3,\ldots, n_i$, form a complete orthonormal set in the separable Hilbert space X.
(3) Operator A generates an analytic semigroup of linear bounded operators $S(t)$: $X \to X$, for $t \geq 0$.

Let $U^+ \subset R^m$ be a positive cone in the space R^m, i.e. $U^+ = \{u \in R^m : u_j \geq 0$ for $j = 1, 2,\ldots, m\}$. We define the set of admissible nonnegative controls U_{ad} as follows

$$U_{ad} = \{u \in L^p_{loc}([0, \infty), R^m) \,;\, u(t) \in U^+ \text{ a.e. on } [0, \infty)\}$$

It is well known (see e.g. [30]), that for each $u \in U_{ad}$ and $x(0) \in X$ there exists unique so called mild solution $x(t, x(0), u) \in D(A)$, $t \geq 0$ of the Eq. (5.15) given by

$$x(t, x(0), u) = S(t)x(0) + \int_0^t S(t - s)Bu(s)ds$$

We say that dynamical system (5.15) is positive if the semigroup S and operator B are positive. In this case the solution $x(t, x(0), u)$ for initial condition $x(0) \in X^+$ and admissible control $u \in U_{ad}$ remains in X^+ for all $t \geq 0$.

We define the attainable or reachable set from the origin in time T by

$$K_T = \{\int_0^T S(T - s)Bu(s)ds \,:\, u \in U_{ad}\}$$

The set

$$K_\infty = \bigcup_{T > 0} K_T$$

is called the attainable or reachable set in finite time.

Using the concept of attainable set we may define different kinds of controllability for dynamical system (5.15). Generally, for infinite dimensional dynamical system it is necessary to introduce two fundamental notions of controllability, namely exact (strong) controllability and approximate (weak) controllability.

However, since our dynamical system has infinite dimensional state space X and finite dimensional control space R^m, then by [30, 31] it is never exactly controllable in any sense. Therefore, in the sequel we shall concentrate only on approximate controllability with positive controls for system (5.15).

Definition 5.3 [30, 31]. Dynamical system (5.15) is said to be approximately controllable with nonnegative controls if $cl\, K_\infty(U^+) = X$.

In the unconstrained case, i.e. when the controls values are taken from the whole space R^m, we say simply about approximate controllability of system (5.15).

The above notion of approximate controllability is defined in the sense that we want to reach a dense subspace of the entire state space. However, in many instances for positive systems with nonnegative controls, it is known that all states are contained in a closed positive cone X^+ of the state space. In this case approximate controllability in the sense of the above definition is impossible but it is interesting to know conditions under which the reachable states are dense in X^+. This observation leads directly to the concept of so-called positive approximate controllability.

Definition 5.4 [30, 31]. Dynamical system (5.15) is said to be approximately positive controllable if $cl\ K_\infty(U^+) = X^+$.

Remark 5.3 From the above two definitions directly follows, that approximate controllability with nonnegative controls always implies approximate positive controllability. However, the converse statement is not generally true.

Finally, we shall recall some fundamental theorems concerning unconstrained and constrained approximate controllability of dynamical system (5.15). In order to do that let us introduce the following notations.

Using eigenvectors x_{ik}, $i = 1, 2, 3, \ldots$ $k = 1, 2, 3, \ldots, n_i$ we introduce for the operator B the following notation:

$$B_i = \left[\langle b_j, x_{ik}\rangle_X\right]_{j=1,2,\ldots m, k=1,2,n_i}$$

B_i, for $i = 1, 2, 3, \ldots$ are $n_i \times m$-dimensional constant matrices which play an important role in controllability investigations [30, 31].

For the case when eigenvalues s_i are simple, i.e. $n_i = 1$, for $i = 1, 2, 3, \ldots$, B_i are m-dimensional row vectors

$$b^i = \left[\langle b_j, x_i\rangle_X\right]_{j=1,2,\ldots,m} \quad \text{for } i = 1, 2, 3, \ldots$$

For simplicity of notation let us denote $b_{ikj} = \langle b_j, x_{ik}\rangle_X$ for $i = 1, 2, 3, \ldots k = 1, 2, \ldots, n_i$, and $j = 1, 2, \ldots, m$. Therefore, we may express matrices B_i and vectors b^i in a more convenient form as follows

$$B_i = \begin{bmatrix} b_{i11} & b_{i11} & \cdots & b_{i11} & \cdots & b_{i11} \\ b_{i11} & b_{i11} & \cdots & b_{i11} & \cdots & b_{i11} \\ \cdots & \cdots & \cdots & \cdots & \cdots & \cdots \\ b_{i11} & b_{i11} & \cdots & b_{i11} & \cdots & b_{i11} \\ \cdots & \cdots & \cdots & \cdots & \cdots & \cdots \\ b_{i11} & b_{i11} & \cdots & b_{i11} & \cdots & b_{i11} \end{bmatrix} \quad \text{for } i = 1, 2, 3, \ldots$$

$$b^i = \begin{bmatrix} b_{i1}, b_{i2}, \ldots, b_{ij}, \ldots, b_{im} \end{bmatrix} \quad \text{for } i = 1, 2, 3, \ldots$$

Since the operator A is selfadjoint, then using the above notations it is possible to express the solution $x(t, x(0), u)$ as follows

5.6 Positive Controllability of Positive Dynamical Systems

$$x(t,x(0),u) = \sum_{i=1}^{i=\infty}\sum_{k=1}^{k=n_i} v_{ik}^0(t)x_{ik} + \sum_{i=1}^{i=\infty}\sum_{k=1}^{k=n_i} v_{ik}^u(t)x_{ik}$$

where

$$v_{ik}^0(t) = \exp(s_i t)\langle x(0), x_{ik}\rangle$$
$$\text{for } i = 1,2,3,\ldots \text{ and } k = 1,2,\ldots,n_i$$

$$v_{ik}^u(t) = \int_0^t \exp s_i(t-\tau)\left(\sum_{j=1}^{j=m} b_{ikj}u_j(\tau)\right)d\tau$$
$$\text{for } i = 1,2,3,\ldots \quad \text{and} \quad k = 1,2,\ldots,n_i$$

We start with the well known (see e.g. [30, 31, 40, 51] for details) necessary and sufficient conditions for approximate controllability with unconstrained controls.

Theorem 5.4 [30, 31]. *Dynamical system (5.15) is approximately controllable if and only if*

$$\text{rank } B_i = n_i \quad \text{for} \quad \text{every } i = 1,2,3,\ldots$$

Corollary 5.5 [30, 31]. *Let m = 1. Then dynamical system (5.15) is approximately controllable 5f and only if every vector $b^i \in R^m$, $i = 1, 2, 3,\ldots$ contains at least one nonzero element.*

Now we recall known in the literature (see [31, 36] for more details) necessary and sufficient condition of approximate controllability with nonnegative controls for dynamical system (5.15).

Theorem 5.5 [30, 36]. *Dynamical system (5.15) is approximately controllable with nonnegative controls if and only if rank $B_i = n_i$ for every $i = 1, 2, 3,\ldots$ and the columns of these matrices B_i, $i = 1,2,3,\ldots$ which correspond to the real eigenvalues, form positive bases in the space R^m.*

Remark 5.4 The above result implies, in particular, that the number of positive controls required for approximate controllability with nonnegative controls is at least that of the highest multiplicity of the eigenvalues plus one. Therefore, dynamical system (5.15) with one scalar nonnegative control is never approximately controllable [30, 31, 36, 66]. Moreover, it should be stressed, that in general case for multiple eigenvalues, it is not so easily to verify the hypothesis that the set of given vectors forms a positive basis in the corresponding Euclidean space.

Remark 5.5 Using the concept of polar cone C^0, the results stated in above theorem can be extended for constrained controls which take their values from a given closed compact cone C with nonempty interior int$C \in U_{ad}$.

Now we shall recall the results concerning approximate positive controllability for dynamical systems (5.15). We start with the following negative result on approximate positive controllability.

Theorem 5.6 [30, 31, 36]. *If there exists p and q such that eigenvalue $s_p \in R$ and coefficients b_{pqj} have the same sign for every $j = 1, 2, \ldots, m$, then the dynamical system (5.15) is not approximately positive controllable.*

From the above theorems and remarks follows the next result concerning approximate controllability of system (5.15) with nonnegative controls.

Corollary 5.6 *If the assumptions of Theorem 5.6 are satisfied, then the dynamical system (5.15) is not approximately controllable with nonnegative controls.*

In the results given above we have obtained some negative results concerning approximate positive controllability for dynamical system (5.15). However, it is often not so important to reach the entire positive cone of the state space. It suffices to steer approximately dynamical system to particular positive states and held constant by a nonnegative control for all times.

This observation directly leads to the concept of so called positive stationary pairs. In this section we generally assume that the dynamical system (5.15) is positive in the sense stated in the next part of the chapter.

Definition 5.5 We call a pair $\{x_s, u_s\} \in (X^+\setminus\{0\}) \times U^+$ positive stationary pair if

$$Ax_s + Bu_s = 0.$$

In this case $x(t, x_s, u_s) = x_s \in X^+$ is a nonzero constant solution of the Eq. (5.15) for $t \geq 0$, $u(t) = u_s$ and $x_s = x(0)$.

Theorem 5.7 [31, 36]. *Let dynamical system (5.15) be positive and S(t) be uniformly exponentially stable positive semigroup. Then to each $u_s \in U^+\setminus kerB$ there exists exactly one $x_s = -A^{-1}Bu_s$ such that $\{x_s, u_s\}$ is a positive stationary pair. Moreover, if $\{x_s, u_s\}$ is a positive stationary pair, and we choose $x(0) \in X^+$ and $u(t) = u_s$, $t \geq 0$, then the solution of the Eq. (5.15) tends to x_s as $t \to \infty$.*

Corollary 5.7 *Let $Re(s_1)<0$. Then to each $u_s \in U^+\setminus kerB$ there exists exactly one*

$$x_s = \sum_{i=1}^{i=\infty} s_i^{-1} \sum_{k=1}^{k=n_i} \left\langle x_{ik}, \sum_{j=1}^{j=m} b_j u_{sj} \right\rangle_X x_{ik} \tag{5.16}$$

such that $\{x_s, u_s\}$ is a positive stationary pair.

Proof Since the spectrum of the linear operator $\sigma(A)$ is pure discrete point spectrum, we conclude that the inequality $Re(s_1)< 0$ is a necessary and sufficient condition for so called uniform stability of linear dynamical system [30]. Therefore, using general spectral formula for the operator A^{-1} and Theorem 5.7 stated above we obtain immediately equality (5.16).

Remark 5.6 Many valuable remarks and comments on the relationships between different kinds of stability (uniform exponential stability, strong stability, weak stability) of the linear abstract differential Eq. (5.15) and the existence of positive stationary pairs for positive dynamical systems can be found in the papers [66, 72].

5.6 Positive Controllability of Positive Dynamical Systems

Now we shall illustrate the general theorems and corollaries stated in the previous sections for the case of linear distributed parameter systems of parabolic type. We begin by describing the mathematical model of the distributed parameter system.

Let Ω be a bounded, open and connected subset of R^N with a smooth boundary $\partial\Omega$ and $cl\Omega = \Omega \cup \partial\Omega$. Let Δ be the Laplacian operator on Ω and ∇ be the gradient operator on Ω. Let us consider linear distributed parameter dynamical system described by the following partial differential equation of parabolic type

$$w_t(z,t) = Aw(z,t) + \sum_{j=1}^{j=m} b_j(z)u_j(t) \quad t>0 \quad z \in \Omega \qquad (5.17)$$

where

$b_j \in L^2(\Omega)$, for $j = 1,2,3,...,m$, and admissible controls are nonnegative, i.e.

$$u_j \in L^2_{loc}([0,\infty), R^+), \quad for \ j = 1,2,3,\ldots,m.$$

The boundary conditions are assumed to be of the following form

$$\alpha(z)w(z,t) + \beta(z)\partial w/\partial v(z,t) = 0 \quad t \geq 0 \quad z \in \partial\Omega \qquad (5.18)$$

It is assumed that $\alpha(z)$ and $\beta(z)$ are twice continuously differentiable on $cl\Omega$. The vector field $v(z)$ is the outer unit normal to $\partial\Omega$ at $z \in \partial\Omega$ and $\partial/\partial v = v\nabla$ denotes differentiation in the direction of the outward norma to Ω. Specifying $\alpha(z)$ and $\beta(z)$ we obtain Dirichlet, Neumann or Robin (mixed) boundary conditions.

The initial condition for Eq. (5.17) is given by

$$w(z,0) = w_0(z) \quad z \in \Omega$$

The second order uniformly elliptic differential operator has the following form

$$A = \sum_{k,j=1}^{k,j=N} a_{kj}(z)D_kD_j + \sum_{k=1}^{k=N} a_k(z)D_k + a_0(z)I \qquad (5.19)$$

where

$z \in R^N$, $a_{kj}(z) = a_{jk}(z)$, for $j,k = 1,2,3,\ldots,N$,
$D_k = \partial/\partial z_k$, for $k = 1,2,3,\ldots,N$.

The domain $D(A)$ of the operator A is characterized explicitly by

$$D(A) = \{w \in L^2(\Omega) : Aw \in L^2(\Omega) \ and \ \alpha(z)w(z,t) + \beta(z)\partial w/\partial v(z,t) = 0, t \geq 0, z \in \partial\Omega\}$$

The coefficients $a_{kj}(z)$, $a_k(z)$ and $a_0(z)$ are assumed to be twice continuously differentiable on Ω and $a_0(z) \geq 0$ for $z \in \Omega$.

Moreover, since operator A is uniformly elliptic then there exists a positive constant μ such that for all vectors $\xi \in R^N$ we have

$$\sum_{k,j=1}^{k,j=N} a_{kj}(z)\xi_k\xi_j \geq \mu|\xi|^2, \quad for\ z \in \Omega$$

Various special cases of (5.17) could be considered, i.e. the reaction diffusion dynamical system

$$w_t(z,t) = d\Delta w(z,t) + aw(z,t) + \sum_{j=1}^{j=m} b_j(z)u_j(t) \quad t > 0 \quad z \in \Omega \quad (5.4)$$

where a and d are real constants.

It is well known (see e.g. [30, 31, 36] for details), that operator A generates an analytic positive semigroup of bounded compact operators $S(t): X \to X$ for $t \geq 0$ [30]. Moreover, since the set Ω is bounded, then the operator A has only pure discrete point spectrum $\sigma(A) = \sigma_p(A) = \{s_1, s_2, s_3, \ldots, s_i, \ldots\}$, consisting entirely with isolated eigenvalues each with finite multiplicity's $n_i < \infty$, $i = 1, 2, 3, \ldots$ and the corresponding set of eigenfunctions $\{x_{ik},\ i = 1, 2, 3, \ldots,\ k = 1, 2, 3, \ldots, n_i\}$ forms an orhonormal basis in the space $L^2(\Omega)$.

An additional property of the operator A that will be important later is stated in the following lemma which is proved in [30].

Lemma 5.1 [30, 31, 36, 66]. *There exists a real eigenvalue s_1 of the operator A and corresponding eigenvector $x_1(z)$ is a strictly positive element in the space X, i.e. satisfies*

$$x_1(z) \gg 0$$

for all $z \in cl\Omega$ in the case of Neumann or Robin (mixed) boundary conditionsand for all $z \in \Omega$ in case of Dirichlet boundary conditions.

In the latter case, we also have

$$\frac{\partial x_1}{\partial v}(z) < 0 \quad for\ z \in \partial\Omega$$

Moreover, if s_i, $i = 2, 3, 4, \ldots$ is any other eigenvalue of the operator A, then the real part of s_i, $Re(s_i)$, satisfies

$$Re(s_i) < s_1 \quad for\ all \quad i = 2, 3, 4, \ldots$$

In other words, Lemma 5.1 says that there exists a real eigenvalue of the operator A which is larger than the real part of all other eigenvalues of the operator

5.6 Positive Controllability of Positive Dynamical Systems

A. We call it the principal eigenvalue of the operator A. Moreover, Lemma 5.1 states, that the associated eigenvector is positive and is called the principal eigenvector of the operator A.

We may express distributed parameter dynamical system (5.17) with boundary conditions (5.18) as an abstract ordinary differential equation in the separable Hilbert space $X = L^2(\Omega)$. Since operator A satisfies all the assumptions stated in the previous parts of the chapter it is sufficient to substitute $x(t) = w(\bullet, t) \in L^2(\Omega) = X$.

Let us denote

$$b_{1j} = \langle b_j, x_1 \rangle_{L^2(\Omega)}$$
$$= \int_\Omega b_j(z) x_1(z) dz \quad for \quad j = 1, 2, 3, \ldots, m$$

Now, using the general results stated before we may formulate theorem and corollaries on positive approximate controllability for distributed parameter dynamical system (5.17) with normal operator A.

Theorem 5.8 *Let operator A be normal. Moreover, let us assume that b_{1j} have the same sign for every $j = 1, 2, \ldots, m$. Then the linear distributed parameter dynamical system (5.17) is not approximately positive controllable.*

Proof Let us observe that distributed parameter dynamical system (5.17) satisfies all the assumptions required in Theorem 5.7. Therefore, by Theorem 5.8 our dynamical system (5.17) is not approximately positive controllable.

Using the results given previously we have the following corollary.

Corollary 5.8 *If $s_1 < 0$, then to each $u_s \in U^+ \backslash kerB$ there exists exactly one x_s such that $\{x_s, u_s\}$ is a positive stationary pair.*

Example 5.2 Let us consider the one dimensional heat equation on a rod of length 1 with noninsulated ends described by the following linear partial differential equation of parabolic type

$$w_t(z,t) = w_{zz}(z,t) + b(z)u(t) \quad 0 \leq z \leq 1 \quad t \geq 0 \quad (5.19)$$

with initial condition

$$w(z, 0) = w_0(z)$$

and Dirichlet type homogeneous boundary conditions

$$w(0, t) = w(1, t) = 0$$

We wish to control distributed parameter system (5.19) by a nonnegative scalar input $u \in L^2_{loc}([0, \infty), R^+)$. In practical applications we can interpret this admissible control as an electrical heating input that for all time is proportional to a given heat distribution $b(z) \in L^2([0, 1], R)$.

We state this infinite dimensional control problem as an abstract control problem defined in the separable Hilbert space $X = L^2([0, 1], R)$. Moreover, let us denote $w(z, t) = x(t) \in X$.

Let $A = d^2/dz^2$ be the linear unbounded selfadjoint differential operator on X with domain $D(A) = \{w(z) \in X: w_{zz}(z) \in X, w(0) = w(1) = 0\}$. It is known, that the operator A has simple eigenvalues $s_i = -i^2\pi^2$ and the corresponding set of eigenfunctions

$$x_i(z) = 2^{0.5} \sin(i\pi z), \text{ for } i = 1, 2, 3, \ldots$$

forms an orthonormal basis in the space $X = L^2([0, 1], R)$.

Since all the eigenvalues are real, then by Theorem 5.7 dynamical systems (5.19) is not approximately positive controllable for any $b \in X$. The same result has been proved in the monographs [30, 31] but using quite different methods.

Let us observe that operator A generates an analytic positive semigroup $S(t)$, for $t \geq 0$ on X given by

$$S(t)x = \sum_{i=1}^{i=\infty} \exp(-i^2\pi^2 t)\langle x, x_i\rangle_{L^2} x_i$$

Now, let us assume that $b \in X^+ = L^2([0, 1], R^+)$. Therefore, distributed parameter system (5.19) is a positive dynamical system. Following [30, 31, 36, 66] it should be stressed, that positive dynamical system (5.19) is also not approximately positive controllable.

However, since $Re(s_1) = -\pi^2 < 0$, then by Corollary 5.8 for each $u_s \in R^+$ there exists exactly one $x_s = -A^{-1}bu_s \in X^+$ given by

$$x_s = \sum_{i=1}^{i=\infty} (-i^2\pi^2)^{-1} \int_0^1 \sqrt{2}\sin(i\pi zz)b(z)dz\sqrt{2}\sin(i\pi zz)u_s$$

such that $\{x_s, u_s\}$ is a positive stationary pair. Using the results stated above an element $x_s \in X^+$ can be also expressed as follows

$$x_s(z) = \left(z\int_0^1\int_0^\xi b(\zeta)d\zeta d\xi - \int_0^z\int_0^\xi b(\zeta)d\zeta d\xi\right)u_s$$

Summarizing, distributed parameter dynamical system (5.19) is not approximately positive controllable and of course it is also not approximately controllable with nonnegative controls, however, for dynamical system (5.19) there exist stationary pairs.

The present paper contains several results on constrained controllability for linear infinite-dimensional selfadjoint dynamical systems. Special attention is paid on positive dynamical systems. Using spectral properties of normal generally

5.6 Positive Controllability of Positive Dynamical Systems

unbounded linear operators with pure discrete point spectrum, conditions for different kinds of constrained controllability have been formulated and proved.

General results have been also applied for constrained controllability considerations for linear distributed parameter dynamical systems described by linear partial differential equations of parabolic type with various kinds of boundary conditions.

Some kinds of the presented results can be extended to cover the case of infinite-dimensional normal dynamical systems with discrete and continuous spectrum. It is also possible to extend the result for second-order infinite-dimensional dynamical systems.

Chapter 6
Constrained Controllability of Second Order Dynamical Systems with Delay

6.1 Introduction

In recent years various controllability problems for different types of nonlinear dynamical systems have been considered in many publications and monographs. The extensive list of these publications can be found for example in the monograph [32] or in the survey paper [33].

However, it should be stressed, that the most literature in this direction has been mainly concerned with controllability problems for finite-dimensional nonlinear dynamical systems with unconstrained controls and without delays or for linear dynamical systems with constrained controls and with delays.

Monograph [14] presents controllability, observability and duality results for continuous and discrete positive linear dynamical systems. Moreover, in the papers [15, 33] controllability and reachability of special kinds of linear stationary control dynamical systems are considered.

In this chapter finite-dimensional dynamical control systems described by second order semilinear stationary ordinary differential state equations with delay in control are considered. Using a generalized open mapping theorem, sufficient conditions for constrained local controllability in a given time interval are formulated and proved. These conditions require verification of constrained global controllability of the associated linear first-order dynamical control system. It is generally assumed, that the values of admissible controls are in a convex and closed cone with vertex at zero.

In the present chapter, we shall consider constrained local controllability problems for second-order finite-dimensional semilinear stationary dynamical systems with point delay in control, described by the set of ordinary differential state equations. Let us recall, that semilinear dynamical control systems contain both linear and pure nonlinear parts in the differential state equations.

In the chapter we shall formulate and prove sufficient conditions for constrained local controllability in a prescribed time interval for semilinear second-order stationary dynamical systems which nonlinear term is continuously differentiable near the origin and with single point delay in control. It is generally assumed that the values of admissible controls are in a given convex and closed cone with vertex at zero, or in a cone with nonempty interior. Proof of the main result is based on the so called generalized open mapping theorem presented in simplified version the paper [36].

Roughly speaking, it will be proved that under suitable assumptions constrained global relative controllability of a linear first-order associated approximated dynamical system implies constrained local relative controllability near the origin of the original semilinear second-order dynamical system. This is a generalization to constrained controllability case some previous results concerning controllability of linear dynamical systems with unconstrained controls [31–33, 36].

Finally, simple numerical example which illustrates theoretical considerations is also given. It should be pointed out, that the results given in the paper extend for the case of semilinear second-order dynamical systems constrained controllability conditions, which were previously known only for linear second-order systems.

6.2 System Description

Let us consider semilinear finite-dimensional control system with single point delay in control described by the following second order differential equation

$$w''(t) = Gw(t) + f(w(t), u(t), u(t-h)) + Hu(t) + Ku(t-h) \quad (6.1)$$
$$\text{for } t \in [0, T]$$

where state vector $w(t) \in R^n = W$ and the control vector $u(t) \in R^m = U$, G is $n \times n$ dimensional constant matrix, H and K are $n \times m$ dimensional constant matrices, $h > 0$ is single point delay.

Moreover, let us assume that nonlinear mapping $f: W \times U \times U \to W$ is continuously differentiable near the origin and such that $f(0, 0, 0) = 0$.

For simplicity of considerations we assume zero initial conditions, i.e.

$$w(0) = 0 \quad \text{and} \quad w'(0) = 0$$

Using standard substitutions

$$x(t) = \begin{bmatrix} x_1(t) \\ x_2(t) \end{bmatrix} = \begin{bmatrix} w(t) \\ w'(t) \end{bmatrix} \in R^{2n}$$

we may transform second-order semilinear dynamical system (6.1) into equivalent first-order semilinear stationary $2n$-dimensional control system described by the following ordinary differential state equation

6.2 System Description

$$x'(t) = Ax(t) + F(x(t), u(t), u(t-h)) + Bu(t) + Du(t-h)$$
$$\text{for } t \in [0, T], \ T > 0 \tag{6.2}$$

with zero initial conditions:

$$x(0) = 0 \ u(t) = 0 \quad \text{for } t \in [-h, 0]$$

where state vector $x(t) \in R^{2n} = X$ and the control $u(t) \in R^{2m} = U$, A is $2n \times 2n$ dimensional constant matrix, B and D are $2n \times m$ dimensional constant matrices.

$$A = \begin{bmatrix} 0 & I \\ G & 0 \end{bmatrix}$$

$$B = \begin{bmatrix} 0 \\ H \end{bmatrix}$$

$$D = \begin{bmatrix} 0 \\ K \end{bmatrix}$$

and the nonlinear term has the form

$$F(x(t), u(t), u(t-h)) = \begin{bmatrix} 0 \\ f(x(t), u(t), u(t-h)) \end{bmatrix} \in R^{2n}$$

Moreover, from previous assumptions concerning nonlinear term $f(x(t), u(t), u(t-h))$ it follows, that the nonlinear mapping $F: X \times U \times U \longrightarrow X$ is also continuously differentiable near the origin and such that $F(0, 0, 0) = 0$.

It is well known, that in practical applications admissible controls are always required to satisfy certain additional constraints. Generally, for arbitrary control constraints it is rather very difficult to give easily computable criteria for constrained controllability even in the linear case and finite dimensional case [31, 33]. However, for some special cases of the constraints it is possible to formulate and prove simple algebraic constrained controllability conditions.

Therefore, in the sequel we shall assume that the set of values of admissible controls $U_c \subset U$ is a given closed and convex cone with nonempty interior and vertex at zero. Then, the set of admissible controls for the dynamical control systems (6.1) and (6.2) has the following form $U_{ad} = L_\infty([0, T], U_c)$.

Then for a given admissible control u(t) there exists a unique solutions $w(t; u) \in R^n$ of the second-order differential Eq. (6.1) and similarly, unique solution $x(t; u) \in R^{2n}$ of the first-order ordinary differential state Eq. (6.2) and with zero initial condition. Transforming semilinear differential Eq. (6.2) into the nonlinear integral equation we have [36],

$$x(t;u) = \int_0^t S(t-s)(F(x(s;u(s)), u(s), u(s-h)) + Bu(s) + Du(s-h))ds \quad (6.3)$$

where the matrix semigroup

$$S(t) = \exp(At) \quad \text{for } t \geq 0$$

is $2n \times 2n$ dimensional exponential transition matrix for the linear part of the semilinear first-order control system (6.2).

For the semilinear stationary finite-dimensional second-order dynamical system (6.1) or equivalently for first-order dynamical system (6.2), it is possible to define many different concepts of controllability. However, in the sequel we shall focus our attention on the so called constrained controllability in a given time interval $[0, T]$.

In order to do that, first of all let us introduce the notion of the so called attainable or reachable set for dynamical system (6.2) at given final time $T > 0$ from zero initial conditions, denoted shortly by $K_T(U_c)$ and defined as follows

$$K_T(U_c) = \{x \in X : x = x(T, u), u(t) \in U_c \quad \text{for a.e. } t \in [0, T]\} \quad (6.4)$$

where $x(t, u)$, $t > 0$ is the unique solution of the differential first-order state Eq. (6.2) with zero initial conditions and a given admissible control $u \in U_{ad} = L_\infty([0, T], U_c)$.

Now, using the concept of the attainable set $K_T(U_c)$, let us recall the well known (see e.g., [31, 32, 36]) definitions of local and global constrained controllability in $[0, T]$ for semilinear second-order dynamical system (6.1).

Definition 6.1 The dynamical system (6.1) is said to be U_c-locally controllable in $[0, T]$ if the attainable set $K_T(U_c)$ contains a neighborhood of zero in the space X.

Definition 6.2 The dynamical system (6.1) is said to be U_c-globally controllable in $[0, T]$ if $K_T(U_c) = X$.

Now, in the last part of this section we shall discuss the relationships between controllability of the first-order system (6.2) for $F(x(t), u(t))=0$, and linear second-order dynamical system (6.1) for the case when $f(w(t), u(t))=0$. Therefore, for comparison we shall consider the following two linear dynamical systems:

$$w''(t) = Gw(t) + Hu(t) + Ku(t-h) \quad t \in [0, T] \quad (6.5)$$

$$x'(t) = Ax(t) + Bu(t) + Du(t-h) \quad t \in [0, T] \quad (6.6)$$

Corollary 6.1 *Second-order linear dynamical system (6.5) is controllable without any control constraints in a given time interval if and only if associated first-order $2n$-dimensional dynamical system (6.6) is controllable without any control constraints in the same given time interval.*

6.2 System Description

Remark 6.1 However, it should be pointed out, that for the controllability problem with constrained controls the above Corollary 6.1 does not hold and there are no general direct relationships between constrained controllability of first-order and second-order linear dynamical systems.

6.3 Controllability Conditions

In order to formulate and prove constrained controllability conditions for system (6.1), general results for nonlinear operator presented in Chap. 5, Sect. 5.3, Lemma 5.1 and Lemma 5.2 will be used.

In this section we shall study constrained local relative controllability in $[0, T]$ for semilinear dynamical system (6.1) using the associated linear $2n$-dimensional control dynamical system

$$z'(t) = Cz(t) + Eu(t) + Gu(t - h) \quad \text{for } t \in [0, T] \tag{6.7}$$

with zero initial condition $z(0) = 0$, where

$$C = A + D_x F(0, 0, 0)$$

$$E = B + D_u F(0, 0, 0)$$

$$G = D + D_{u(t-h)} F(0, 0, 0) \tag{6.8}$$

The main result of the paper is the following sufficient condition for constrained local controllability in a given time interval of the semilinear dynamical system with single point delay in control (6.1).

Theorem 6.1 *Suppose that*

(i) $F(0, 0, 0) = 0$,
(ii) $U_c \subset U$ is a closed and convex cone with vertex at zero,
(iii) *The associated linear control system* (6.7) *is U_c-globally controllable in* $[0, T]$.

Then the semilinear stationary dynamical control system (6.1) *is U_c-locally controllable in* $[0, T]$.

Proof Let us define for the semilinear dynamical system (6.1) a nonlinear map

$$g : L_\infty([0, T], U_c) \to X \text{ by } g(u) = x(T, u).$$

Similarly, for the associated linear dynamical system (6.7), we define a linear map

$$H : L_\infty([0, T], U_c) \to X \text{ by } Hv = z(T, v).$$

By the assumption (iii) the linear dynamical system (6.7) is U_c-globally relative controllable in [0, T]. Therefore, by the Definition 6.2 the linear operator H is surjective, i.e. it maps the convex cone U_{ad} onto the whole state space X. Furthermore, by Lemma 5.1 and Lemma 5.2 we have that $Dg(0) = H$.

Since U_c is a closed and convex cone, then the set of admissible controls $U_{ad} = L_\infty([0, T], U_c)$ is also a closed and convex cone in the function space $L_\infty([0, T], U)$. Therefore, the nonlinear map g satisfies all the assumptions of the generalized open mapping theorem stated in the Lemma 5.1.

Hence, the nonlinear map g transforms a conical neighborhood of zero in the set of admissible controls U_{ad} onto some neighborhood of zero in the state space X. This is by Definition 6.1 equivalent to the U_c-local relative controllability in [0, T] of the semilinear dynamical control system (6.1). Hence, our theorem follows.

In practical applications of the Theorem 6.1, the most difficult problem is to verify the assumption (iii) about constrained global controllability of the linear stationary dynamical system (6.1). In order to avoid this disadvantage, we may use the following Theorem.

Theorem 6.2 [31–33, 36]. Suppose that the set U_c is a given convex cone with vertex at zero and a nonempty interior in the space of control values R^m. Then the associated linear dynamical control system with single point delay in control (6.1) is U_c-globally controllable in given time interval [0, T] for $T \leq h$ if and only if

(i) it is controllable without any constraints, i.e.,

$$\text{rank}\left[E, CE, C^2E, \ldots, C^{2n-1}E\right] = 2n,$$

(ii) there is no real eigenvector $v \in R^{2n}$ of the matrix C^{tr} satisfying inequalities

$$v^{tr}Eu \leq 0, \text{ for all } u \in U_c.$$

Moreover, the associated linear dynamical control system (6.7) is U_c-globally controllable in [0, T] for $T > h$ if and only if

(iii) it is controllable without any constraints, i.e.

$$\text{rank}\left[E, G, CE, CG, C^2E, C^2G, \ldots, C^{2n-1}E, C^{2n-1}G\right] = 2n,$$

(iv) there is no real eigenvector $v \in R^{2n}$ of the matrix C^{tr} satisfying inequalities

$$v^{tr}Du \leq 0, \text{ for all } u \in U_c.$$

6.3 Controllability Conditions

It should be pointed out that for the single input associated linear dynamical control system (6.7), i.e. for the case of scalar controls and $m = 1$, Theorem 6.2 reduces to the following Corollary.

Corollary 6.1 *Suppose that the dynamical system* (6.1) *has single input, i.e.* $m = 1$ *and* $U_c = R^+$.

Then the associated linear dynamical control system (6.7) is U_c-globally controllable in $[0, T]$, for $T \leq h$ if and only if it is controllable without any constraints, i.e.

$$rank\left[E, CE, C^2E, \ldots, C^{2n-1}E\right] = 2n,$$

and matrix C has only complex eigenvalues.

Moreover, the associated linear dynamical control system with single point delay in control (6.7) is U_c-globally controllable in given time interval $[0, T]$, for $T > h$ if and only if it is controllable without any constraints, i.e.

$$rank\left[E, G, CE, CG, C^2E, CG^2, \ldots, C^{2n-1}E, C^{2n-1}G\right] = 2n,$$

and matrix C has only complex eigenvalues.

Remark 6.2 It should be stressed that the important advantage of the Corollary 6.1 is that instead rather difficult condition (ii) given in Theorem 6.2 it is enough to verify only eigenvalues of the matrix C.

Remark 6.3 Since all the dynamical systems considered in the previous sections are system with constant coefficient and control values are restricted only in direction and are not restricted in their values, then in fact all the results presented in Sect. 6.4 are valid for any time interval $[0, T]$.

Remark 6.4 For dynamical systems with delays, controllability strongly depends on the length of the time interval $[0, T]$. It is well known [30] that dynamical system with delay $h > 0$ may be uncontrollable for $T \leq h$ however, this system may be controllable for the final time $T > h$.

Remark 6.5 General assumption that all initial conditions given in Sect. 6.2 are zero is not essential for controllability considerations for linear dynamical systems with cone constrained values of controls. It should be pointed out, that the same controllability conditions hold for any nonzero initial conditions.

Example 6.1 Finally, let us consider constrained controllability of the simple illustrative example of dynamical systems presented in the previous sections.

Let the semilinear second-order finite-dimensional dynamical control system with point delay in control defined on a given time interval $[0, T]$, has the following form

$$\begin{aligned} w_1''(t) &= -w_1(t) + u(t-h) + e^{u(t)} - 1 \\ w_2''(t) &= -2w_2(t) + \sin w_2(t) + u(t) \end{aligned} \tag{6.9}$$

Therefore, taking into account the previous notations and equations we have $n = 2$, $m = 1$, $w(t) = (w_1(t), w_2(t))^{tr} \in R^2 = W$, $u(t) \in U_c = R^+$.

Now, let us introduce the following standard substitution for state variables:

$$x(t) = \begin{bmatrix} x_1(t) \\ x_2(t) \\ x_3(t) \\ x_4(t) \end{bmatrix} = \begin{bmatrix} w_1(t) \\ w_1'(t) \\ w_2(t) \\ w_2'(t) \end{bmatrix}$$

Then using the notations given in the previous sections matrices A, B, C and D and the nonlinear mapping f have the following form

$$A = \begin{bmatrix} 0 & 1 & 0 & 0 \\ -1 & 0 & 0 & 0 \\ 0 & 0 & 0 & 1 \\ 0 & 0 & -2 & 0 \end{bmatrix}$$

$$B = \begin{bmatrix} 0 \\ 0 \\ 0 \\ 1 \end{bmatrix} \quad D = \begin{bmatrix} 0 \\ 1 \\ 0 \\ 0 \end{bmatrix}$$

$$f(w(t), u(t), u(t-h)) = \begin{bmatrix} e^{u(t-h)} - 1 \\ \sin w_2(t) \end{bmatrix}$$

Therefore, taking into account the form of the Eq. (6.9) we have

$$F(w(t), u(t), u(t-h)) = \begin{bmatrix} 0 \\ e^{u(t-h)} - 1 \\ 0 \\ \sin w_2(t) \end{bmatrix}$$

Moreover, let the cone of values of controls be a cone of positive numbers, i.e. $U_c = R^+$, and therefore, the set of admissible controls has is a cone of the following form $U_{ad} = L_\infty([0, T], R^+)$.

Hence, we have

$$F(0, 0, 0) = \begin{bmatrix} 0 \\ 0 \\ 0 \\ 0 \end{bmatrix}$$

6.3 Controllability Conditions

Moreover,

$$D_x F(x(t), u(t), u(t-h)) = \begin{bmatrix} 0 & 0 & 0 & 0 \\ 0 & 0 & 0 & 0 \\ 0 & 0 & 0 & 0 \\ 0 & 0 & \cos w_2(t) & 0 \end{bmatrix}$$

Therefore,

$$D_x F(0, 0, 0) = \begin{bmatrix} 0 & 0 & 0 & 0 \\ 0 & 0 & 0 & 0 \\ 0 & 0 & 0 & 0 \\ 0 & 0 & 1 & 0 \end{bmatrix}$$

and consequently we have

$$C = A + D_x F(0, 0, 0) = \begin{bmatrix} 0 & 1 & 0 & 0 \\ -1 & 0 & 0 & 0 \\ 0 & 0 & 0 & 1 \\ 0 & 0 & -1 & 0 \end{bmatrix}$$

Similarly,

$$D_u F(x(t), u(t), u(t-h)) = \begin{bmatrix} 0 \\ 0 \\ 0 \\ 0 \end{bmatrix}$$

$$E = B + D_u F(0, 0, 0) = B = \begin{bmatrix} 0 \\ 0 \\ 0 \\ 1 \end{bmatrix}$$

and finally,

$$D_{u(t-h)} F(x(t), u(t), u(t-h)) = \begin{bmatrix} 0 \\ e^{u(t-h)} \\ 0 \\ 0 \end{bmatrix}$$

$$D_{u(t-h)} F(0, 0, 0) = \begin{bmatrix} 0 \\ 1 \\ 0 \\ 0 \end{bmatrix}$$

Hence,

$$G = D + D_{u(t-h)}F(0,0,0) = \begin{bmatrix} 0 \\ 1 \\ 0 \\ 1 \end{bmatrix}$$

Thus, we have

$$C^2 = \begin{bmatrix} -1 & 0 & 0 & 0 \\ 0 & -1 & 0 & 0 \\ 0 & 0 & -1 & 0 \\ 0 & 0 & 0 & -1 \end{bmatrix}$$

$$C^3 = \begin{bmatrix} 0 & -1 & 0 & 0 \\ 1 & 0 & 0 & 0 \\ 0 & 0 & 0 & -1 \\ 0 & 0 & 1 & 0 \end{bmatrix}$$

Moreover,

$$sI - C = \begin{bmatrix} s & -1 & 0 & 0 \\ 1 & s & 0 & 0 \\ 0 & 0 & s & -1 \\ 0 & 0 & 1 & s \end{bmatrix}$$

Therefore, the characteristic equation for the matrix C is as follows

$$\det(sI - C) = (s^2 + 1)(s^2 + 1) = 0$$

and hence, the matrix C has only two different complex eigenvalues i and $-i$ each of multiplicity 2.

Moreover, using well known controllability matrix and rank controllability condition for linear first order dynamical system for time interval $[0, T]$, $T \leq h$, [31–33, 36] we have

$$\text{rank}[E, CE, C^2E, C^3E, C^3] = \text{rank}\begin{bmatrix} 0 & 0 & 0 & 0 \\ 0 & 0 & 0 & 0 \\ 0 & 0 & 0 & -1 \\ 1 & 0 & -1 & 0 \end{bmatrix} = 3 < 4 = 2n$$

Thus, the associated first order linear dynamical system with point delay in control is not controllable in the time interval $[0, T]$ for $T \leq h$.

However, using well known controllability matrix and rank controllability condition for linear dynamical system (6.7) for time interval $[0, T]$, $T > h$ [31–33, 36],

6.3 Controllability Conditions

$$rank[E, G, CE, CG, C^2E, C^2G, C^3E, C^3G] =$$

$$= \begin{bmatrix} 0 & 0 & 0 & 1 & 0 & 0 & 0 & -1 \\ 0 & 1 & 0 & 0 & 0 & -1 & 0 & 0 \\ 0 & 0 & 1 & 1 & 0 & 0 & -1 & 0 \\ 1 & 1 & 0 & 0 & -1 & -1 & 0 & 0 \end{bmatrix} = 4 = 2n \quad (6.10)$$

Hence, both assumptions of the Theorem 6.2 are satisfied and therefore, the associated linear $2n$-dimensional dynamical control system (6.7) with above matrices C, E and G is R^+-globally controllable in a given time interval $[0, T]$, for $T > h$.

Moreover, all the assumptions stated in the Theorem 6.2 are also satisfied and thus, the second-order semilinear dynamical control systems with point delay in control (6.9) is R^+-locally controllable in $[0, T]$. This example shows, that controllability strongly depends on time interval and delayed control.

In this Chapter sufficient conditions for constrained local controllability near the origin for semilinear second-order stationary finite-dimensional dynamical control systems with point delay in control have been formulated and proved. It was generally assumed, that control values are in a given convex cone with vertex at zero and nonempty interior. In the proof of the main result generalized open mapping theorem has been used.

These conditions extend to the case of constrained controllability of second-order finite-dimensional semilinear dynamical control systems with point delay in control the results published previously in [31–33, 36].

The method presented in the Chapter is quite general and covers wide class of semilinear dynamical control systems. Therefore, similar constrained controllability results may be derived for more general class of semilinear dynamical control systems.

For example, it seams, that it is possible to extend sufficient constrained controllability conditions given in the previous sections for more general class of semilinear dynamical control systems with multiple point delay in the control or with multiple point delays in the controls and in the state variables and for the discrete-time semilinear control systems.

Moreover, quite similar method can be used to derive sufficient conditions for local controllability of semilinear dynamical systems with nonlinear term containing both state variables and control function.

6.4 Constrained Controllability of Second Order Infinite Dimensional Systems

In the present chapter constrained approximate controllability of linear abstract second-order infinite-dimensional dynamical systems is considered. It is proved using the frequency-domain method, that constrained approximate controllability of second-order system can be verified by the constrained approximate controllability conditions for the simplified suitable defined first-order system.

General results are then applied for constrained approximate controllability investigation of mechanical flexible structure vibratory dynamical system with damping. For such dynamical systems direct verification of constrained approximate controllability is rather difficult and complicated [53, 57]. Therefore, using frequency-domain method [57] it is shown that constrained approximate controllability of second order dynamical system can be checked by the constrained approximate controllability condition for suitable defined simplified first order dynamical system.

Let V and U denote separable Hilbert spaces. If set $M \subseteq V$, we define the polar cone by $M^o = \{w \in V, \langle w, v \rangle_V \leq 0 \text{ for all } v \in M\}$. The closure, the convex hull and the interior are denoted respectively by cl M, co M and int M. The linear subspace spanned by M is denoted by span M.

Let $A: V \supset D(A) \longrightarrow V$ be a linear generally unbounded self-adjoint and positive-definite linear operator with dense domain $D(A)$ in V and compact resolvent $R(s; A)$ for all s in the resolvent set $\rho(A)$. Then operator A has the following properties [31]:

(1) Operator A has only pure discrete point spectrum $\sigma_p(A)$ consisting entirely with isolated real positive eigenvalues

$$0 < s_1 < s_2 < \ldots < s_i < \ldots, \quad \lim_{i \to \infty} s_i = +\infty$$

Each eigenvalue s_i has finite multiplicity $n_i < \infty$, $i = 1, 2, 3, \ldots$ equal to the dimensionality of the corresponding eigenmanifold.

(2) The eigenvectors $v_{ik} \in D(A)$, $i = 1,2,3, \ldots$ $k = 1, 2, 3, \ldots, n_i$, form a complete orthonormal set in the separable Hilbert space V.

(3) A has spectral representation

$$Av = \sum_{i=1}^{i=\infty} s_i \sum_{k=1}^{k=n_i} \langle v, v_{ik} \rangle_V v_{ik} \quad for \quad v \in D(A)$$

(4) Fractional powers A^α, $0 < \alpha \leq 1$ of the operator A can be defined as follows

$$A^\alpha v = \sum_{i=1}^{i=\infty} s_i^\alpha \sum_{k=1}^{k=n_i} \langle v, v_{ik} \rangle_V v_{ik} \quad for \quad v \in D(A^\alpha)$$

$$\text{where} \quad D(A^\alpha) = \left\{ v \in V : \sum_{i=1}^{i=\infty} s_i^{2\alpha} \sum_{k=1}^{k=n_i} |\langle v, v_{ik} \rangle_V|^2 \langle \infty \right\}$$

(5) Operators A^α, $0 < \alpha \leq 1$ are self-adjoint, positive-definite with dense domains in V and generate analytic semigroups on V.

6.4 Constrained Controllability of Second Order Infinite Dimensional Systems

Let us consider linear infinite-dimensional control system described by the following abstract second order differential equation

$$\ddot{v}(t) + 2\left(c_2 A + c_1 A^{1/2}\right)\dot{v}(t) + \left(d_2 A + d_1 A^{1/2}\right)v(t) = Bu(t) \tag{6.11}$$

where $c_1 \geq 0$, $c_2 \geq 0$, d_1 unrestricted in sign, $d_2 > 0$ are real given constants.

It is assumed, that the operator $B{:}U \longrightarrow V$ is linear operator and its adjoint operator $B^*: V \longrightarrow U$ is $A^{1/2}$-bounded, [31, 53, 57], i.e. $D(B^*) \supset D(A^{1/2})$ and there is positive real number M such that

$$\|B^* v\|_U \leq M\left(\|v\|_v + \|A^{1/2} v\|_v\right) \quad \text{for} \quad v \in D(A^{1/2})$$

Let $\Omega \subset U$ be a convex cone with vertex at the origin in U such that int co $\Omega \neq \emptyset$. In the sequel it is generally assumed, that the admissible controls $u \in L^2_{loc}([0, \infty), \Omega)$.

It is well known [30, 31, 57], that abstract ordinary differential Eq. (6.11) with initial conditions $v(0) \in D(A)$, $v'(0) \in V$ has for each $t_1 > 0$ an unique solution

$$v(t; v(0), v'(0), u) \in C^{(2)}([0, t_1], V)$$

such that

$$v(t) \in D(A) \text{ and } v'(t) \in D(A), \quad \text{for } t \in (0, t_1].$$

Moreover, for $v(0) \in V$ there exists so called "mild solution" for the Eq. (6.15) in the product space $W = V \times V$ with inner product defined as follows

$$\langle v, w \rangle_W = \langle v_1, w_1 \rangle_V + \langle v_2, w_2 \rangle_V$$

In order to transform second order Eq. (6.11) into the first order equation in the Hilbert space W let us substitute [30, 31, 57]:

$$v(t) = w_1(t), \quad v'(t) = w_2(t)$$

Then Eq. (6.11) becomes

$$w'(t) = Fw(t) + Gu(t) \tag{6.12}$$

where

$$w(t) = \begin{bmatrix} w_1(t) \\ w_2(t) \end{bmatrix}, \quad F = \begin{bmatrix} 0 & I \\ d_2 A + d_1 A^{1/2} & c_2 A + c_1 A^{1/2} \end{bmatrix}, \quad G = \begin{bmatrix} 0 \\ B \end{bmatrix}$$

Since the operators A and $A^{1/2}$ are self-adjoint we can obtain for the operator F its adjoint operator F^* as follows

$$F^* = \begin{bmatrix} 0 & d_2 A^* + d_1 A^{1/2*} \\ I & 2(c_2 A^* + c_1 A^{1/2*}) \end{bmatrix} = \begin{bmatrix} 0 & d_2 A + d_1 A^{1/2} \\ I & 2(c_2 A + c_1 A^{1/2}) \end{bmatrix}$$

Remark 6.6 It should be pointed out, that if $c_1^2 + c_2^2 > 0$, then operators F and F^* generate analytic semigroups of linear bounded operators on the Hilbert space $V \times V$ [31, 57]. However, for the case when $c_1 = c_2 = 0$ operator F generates a group of linear bounded operators, which unfortunately cannot be analytic since linear operator F is unbounded [19]. This statements are important for controllability investigations.

In the sequel, for comparison we shall consider instead of the second-order Eqs. (6.11) or (6.12) also simplified first-order differential equation

$$\dot{v}(t) = -A^\alpha v(t) + Bu(t) \quad \text{for } 0 < \alpha < \infty \tag{6.13}$$

In the next parts of the present section we shall also consider second order dynamical system (6.11) with finite-dimensional values of admissible controls, i.e. space $U = R^m$. In this special case, for convenience we shall introduce the following notations:

$$B = [b_1, b_2, \ldots, b_j, \ldots, b_m] \quad \text{where} \quad b_j \in V, \text{for} \quad j = 1, 2, 3, \ldots, m$$
$$u(t) = [u_1(t), u_2(t), u_3(t), \ldots, u_j(t), \ldots, u_m(t)]^T \quad \text{where} \quad u(t) \in L^2_{loc}([0, \infty), \Omega)$$

Let us observe, that in this special case linear operator B is finite-dimensional and therefore, it is a compact operator [31, 57, 72].

Using eigenvectors v_{ik}, $i = 1, 2, 3, \ldots$ $k = 1, 2, 3, \ldots, n_i$ we introduce for finite-dimensional operator B the following notation [31, 57];

$$B_i = \begin{bmatrix} \langle b_1, v_{i1} \rangle_V & \langle b_2, v_{i1} \rangle_V & \cdots & \langle b_j, v_{i1} \rangle_V & \cdots & \langle b_m, v_{i1} \rangle_V \\ \langle b_1, v_{i2} \rangle_V & \langle b_1, v_{i2} \rangle_V & \cdots & \langle b_j, v_{i2} \rangle_V & \cdots & \langle b_m, v_{i2} \rangle_V \\ \cdots & \cdots & \cdots & \cdots & \cdots & \cdots \\ \langle b_1, v_{ik} \rangle_V & \langle b_2, v_{ik} \rangle_V & \cdots & \langle b_j, v_{ik} \rangle_V & \cdots & \langle b_m, v_{ik} \rangle_V \\ \cdots & \cdots & \cdots & \cdots & \cdots & \cdots \\ \langle b_1, v_{im_i} \rangle_V & \langle b_2, v_{im_i} \rangle_V & \cdots & \langle b_j, v_{i1} \rangle_V & \cdots & \langle b_m, v_{im_i} \rangle_V \end{bmatrix} \tag{6.14}$$

for $i = 1, 2, 3, \ldots$

B_i, for $i = 1, 2, 3, \ldots$ are $n_i \times m$-dimensional constant matrices which play an important role in controllability investigations [31, 57]. For the case when eigenvalues s_i are simple, i.e. $n_i = 1$, for $i = 1, 2, 3, \ldots$, B_i are m-dimensional row vectors.

For infinite-dimensional dynamical systems we may introduce two general kinds of controllability, i.e. approximate (weak) controllability and exact (strong) controllability [30, 31]. However, it should be mentioned, that in the case when the semigroup associated with the dynamical system is compact or the control operator is compact, then dynamical system is never exactly controllable in infinite-dimensional

6.4 Constrained Controllability of Second Order Infinite Dimensional Systems

state space [31]. Therefore, in the present paper we shall concentrate on constrained approximate controllability for second order dynamical system (6.11) or equivalently (6.12).

Definition 6.1 [30, 31, 57]. Dynamical system (2.1) is said to be Ω-approximately controllable if for any initial condition $w(0) \in V \times V$, any given final condition $w_f \in V \times V$ and each positive real number ε there exist a finite time $t_1 < \infty$ (depending generally on $w(0)$ and w_f) and an admissible control $u \in L^2([0, t_1], \Omega)$ such that

$$\|W(t_1; W(0), u) - W_f\|_{V \times V} \leq \varepsilon$$

Now, let us recall several well known lemmas [31, 57], concerning constrained approximate controllability of first order dynamical system (2.2), which will be useful in the sequel.

Lemma 6.1 [57]. *Dynamical system (6.12) is U-approximately controllable if and only if for any complex number z, there exists no nonzero $w \in D(F^*)$ such that*

$$\begin{bmatrix} F^* - z \\ G^* \end{bmatrix} w = 0 \tag{6.15}$$

Similarly, dynamical system (6.13) is U-approximately controllable if and only if any complex number s there exists no nonzero $v \in D(A^\alpha) \subset V$ such that

$$\begin{bmatrix} A^{1/2} - sI \\ B^* \end{bmatrix} z = 0$$

Lemma 6.2 [57]. *Suppose that $U = R^m$ and $\Omega = \{u \in R^m = U: u_j(t) \geq 0 \text{ for } t\ 0\}$, then dynamical system (6.17) is Ω-approximately controllable if and only if the columns of the matrices B_i form a positive basis in the space R^{n_i} for every $i = 1, 2, 3, \ldots$*

Lemma 6.3 [57]. *Dynamical system (6.13) is U-approximately controllable if and only if it is approximately controllable for some $\alpha \in (0, \infty)$.*

Lemma 6.4 [72]. *Dynamical system (6.12) is Ω-approximately controllable if and only if it is U-approximately controllable and*

$$\text{Ker}(zI - F) \cap (G\Omega)^o = \{0\} \quad \text{for every } z \in R \tag{6.16}$$

Remark 6.6 Since operator A is selfadjoint then from the above Lemmas directly follows that the dynamical system (6.13) is Ω-approximately controllable if and only if

$$\text{Ker}(sI - A^\alpha) \cap (B\Omega)^o = \{0\} \quad \text{for every } s \in R \tag{6.17}$$

Proposition 6.1 *Dynamical system (6.13) is Ω-approximately controllable if and only if it is Ω-approximately controllable for some $\alpha \in (0, \infty)$.*

Proof Since operator A is selfadjoint and positive definite, then for $\alpha \in (0, \infty)$ we have

$$Ker(sI - A) = Ker(s^{\alpha}I - A^{\alpha}) = Ker(zI - A^{\alpha}),$$

where $z = s^{\alpha}$ is a homeomorphizm.

Hence our proposition follows.

Now, using the frequency-domain method [31] we shall formulate and prove necessary and sufficient condition for approximate controllability for dynamical system (6.11), which is the main result of the present paper.

Theorem 6.3 *Dynamical system (6.11) is Ω-approximately controllable if and only if dynamical system (6.13) is Ω-approximately controllable for some $\alpha \in (0, \infty)$.*

Proof By Proposition 6.1 in order to prove Theorem 6.3 it is sufficient to show the equivalence of the conditions (6.15), (6.16) and equality (6.17) for some α $(0,\infty)$. Therefore, for simplicity of considerations we shall take $\alpha = \frac{1}{2}$.

Now, we shall prove the implication

(6.15) and (6.16) $\Longrightarrow$ (6.17).

By contradiction. Let us suppose, that for some $s \in R$ there exists a nonzero $v \in D(A^{\frac{1}{2}})$ satisfying equality (6.17), i.e.

$$A^{1/2}v = sv \quad \text{and} \quad v \in (B\Omega)^o \qquad (6.18)$$

Thus, s is an eigenvalue of positive-definite operator $A^{\frac{1}{2}}$. Hence, s is real and positive. The vector $v \neq 0$ is the associated eigenvector. Taking into account the form of the linear operators F and G, the equality (6.17) produce the following set of relations

$$d_2 A w_2 + d_1 A^{1/2} w_2 = z w_1 \qquad (6.19)$$

$$w_1 + 2\left(c_2 A + c_1 A^{1/2}\right) w_2 - z w_2 = 0 \qquad (6.20)$$

$$w_2 \in (B\Omega)^o \qquad (6.21)$$

Hence, for $w_2 = v \neq 0$, from (6.17) and (6.19) it follows

$$w_1 = -2\left(c_2 A + c_1 A^{1/2}\right) v - zv = -2\left(c_2 s^2 + c_1 s\right) v - zv \qquad (6.22)$$

Substituting (6.21) into (6.18) and taking into account (6.17) yields

$$(d_2 s^2 + d_1 s) v = -2(c_2 s^2 + c_1 s) zv - z^2 v$$

6.4 Constrained Controllability of Second Order Infinite Dimensional Systems

Thus we have following second order algebraic equation

$$z^2 + 2(c_2s^2 + c_1s)z + (d_2s^2 + d_1s) = 0 \tag{6.23}$$

Its solutions $z_{1,2}$ are given by the equalities

$$z_1 = -2(c_2s^2 + c_1s) - \Delta$$

$$z_2 = -2(c_2s^2 + c_1s) + \Delta$$

where $\Delta = 2\left((c_2s^2 + c_1s)^2 - (d_2s^2 + d_1s)\right)^{1/2}$

Thus, the nonzero vector

$$w = (w_1, w_2) = \left(2(d_2s^2A + d_1sA^{1/2})v + z_{1,2}v, v\right) \in W$$

satisfies (6.19, (6.20) and 6.21). This provides the contradiction.

Now, we shall prove the implication

(6.17) $\Longrightarrow$ (6.15) and (6.16)

By contradiction. Suppose that for some z there exists a nonzero vector $w = (w_1, w_2) \in D(F^*)$ satisfying equalities (6.15) or (6.16), i.e.

$$\left(d_2A + d_1A^{1/2}\right)w_2 = zw_1 \tag{6.24}$$

$$w_1 + 2\left(c_2A + c_1A^{1/2}\right)w_2 = zw_2 \tag{6.25}$$

$$B^*w_2 = 0 \quad \text{or} \quad w_2 \in (B\Omega)^o \quad \text{if } z \in R \tag{6.26}$$

From (6.25) directly follows, that for $w_2 = 0$ we have $w_1 = 0$. Thus, $w_2 \neq 0$ and hence we can take $v = w_2$.

Therefore,

$$w_1 = z^{-1}\left(d_2A + d_1A^{1/2}\right)w_2 = z^{-1}\left(d_2A + d_1A^{1/2}\right)v \tag{6.27}$$

Substituting equality (6.27) into (6.25) we have

$$z^{-1}\left(d_2A + d_1A^{1/2}\right)v + 2\left(c_2A + c_1A^{1/2}\right)v = zv \tag{6.28}$$

Hence,

$$z^2v - 2z\left(c_2A + c_1A^{1/2}\right)v - \left(d_2A + d_1A^{1/2}\right)v = 0 \tag{6.29}$$

Therefore,
$$(-2zc_2 - d_2)Av + (-2zc_1 - d_1)A^{1/2}v + z^2 v = 0 \qquad (6.30)$$

In order to solve the Eq. (6.30) with respect to $A^{1/2}$, let us consider the following three cases:

1. If $2zc_2 + d_2 = 0$ and $2zc_1 + d_1 \neq 0$, then by (6.30) we have
$$A^{1/2}v = z^2(2zc_1 + d_1)^{-1} v \qquad (6.31)$$

2. If $2zc_2 + d_2 \neq 0$ and $2zc_1 + d_1 = 0$, then by (6.30) we have
$$Av = z^2(2zc_2 + d_2)^{-1} v$$

or, equivalently
$$A^{1/2}v = z(2zc_2 + d_2)^{-1/2} v \qquad (6.32)$$

3. If $2zc_2 + d_2 \neq 0$ and $2zc_1 + d_1 \neq 0$, then solving (6.30) as a second order algebraic equation with respect to $A^{1/2}$ we have
$$A^{1/2}v = s_1 v \quad \text{or} \quad A^{1/2}v = s_2 v \qquad (6.33)$$

where
$$s_1 = 0,5(2zc_2 + d_2)^{-1}(2zc_1 + d_1 + \Delta)$$
$$s_2 = 0,5(2zc_2 + d_2)^{-1}(2zc_1 + d_1 - \Delta)$$
$$\Delta = \left[(2zc_1 + d_1)^2 + 4z^2(2zc_2 + d_2)\right]^{1/2}$$

Therefore, equalities (6.31), (6.32) (6.33) together with the inclusion $w_2 \in (B\Omega)^o$ imply that there exists $v \neq 0$ satisfying (6.17) which provides contradiction. Hence Theorem 6.3 follows.

Corollary 6.2 *Suppose that* $\Omega = \{u \in R^m = U: u_j(t) \geq 0 \text{ for } t \geq 0\}$. *Then dynamical system* (6.11) *is* Ω*-approximately controllable if and only if columns of the matrices* B_i *form a positive basis in the space* R^{n_i} *for every* $i = 1, 2, 3, \ldots$

Proof If the columns of the matrices B_i form a positive basis in the space R^{n_i} for every $i = 1, 2, 3, \ldots$ and Ω is a positive cone in the space R^m, then image $B\Omega$ is the whole space R^{n_i} for every $i = 1, 2, 3, \ldots$ Therefore, our Corollary 3.1 follows.

Corollary 6.3 *Suppose that* $c_1^2 + c_2^2 > 0$ *and* $\Omega = U$. *Then dynamical system* (6.11) *is U-approximately controllable in any time interval* $[0, t_1]$ *if and only if dynamical system* (6.12) *is U-approximately controllable in finite time.*

6.4 Constrained Controllability of Second Order Infinite Dimensional Systems

Proof Since for the case when $c_1^2 + c_2^2 > 0$ operator F generates analytic semigroup then approximate controllability of dynamical system (6.12) and hence also of dynamical system (6.11) is equivalent to its approximate controllability in any time interval $[0, t_1]$ [31, 57]. Therefore, from Theorem 3.1 immediately follows Corollary 6.3.

Corollary 6.4 Suppose that $c_1^2 + c_2^2 > 0$, $\Omega = U$, and the space of control values is finite-dimensional, i.e. $U = R^m$. Then the dynamical system (6.11) is U-approximately controllable in any time interval $[0, t_1]$ if and only if

$$\text{rank } B_i = n_i \quad \text{for } i = 1, 2, 3, \ldots \quad (6.34)$$

Proof Corollary 6.4 is a direct consequence of the Theorem 6.3, Corollary 6.3 and well known results [57, 72], concerning approximate controllability of infinite-dimensional dynamical systems with finite-dimensional controls.

Corollary 6.5 Suppose that $c_1^2 + c_2^2 > 0$, $\Omega = U$, the space of control values is finite-dimensional, i.e. $U = R^m$, and moreover, multiplicities $n_i = 1$ for $i = 1, 2, 3, \ldots$ Then dynamical system (6.11) is U-approximately controllable in any time interval $[0, t_1]$ if and only if

$$\sum_{j=1}^{j=m} \langle b_j, v_i \rangle_V^2 \neq 0 \quad \text{for } i = 1, 2, 3, \ldots \quad (6.35)$$

Proof From Corollary 6.4 immediately follows that for the case when multiplicities $n_i = 1$ for $i = 1, 2, 3, \ldots$ dynamical system (6.11) is U-approximately controllable in any time interval if and only if m-dimensional row vectors

$$B_i = \begin{bmatrix} \langle b_1, v_i \rangle_V & \langle b_2, v_i \rangle_V & \cdots & \langle b_j, v_i \rangle_V & \cdots & \langle b_m, v_i \rangle_V \end{bmatrix} \neq 0 \quad \text{for } i = 1, 2, 3, \quad (6.36)$$

Thus, Corollary 6.5 follows.

At the end of this section we shall consider a vibratory dynamical system described by the following linear partial differential equation [57]

$$w_{tt}(t,x) + 2c_1 w_{txx}(t,x) + 2c_2 w_{txxxx}(t,x) + d_1 w_{xx}(t,x) + d_2 w_{xxxx}(t,x)$$
$$= \sum_{j=1}^{j=r} b_j(x) u_j(t) \quad (6.37)$$

defined for $x \in [0, L]$ and $t \in [0, \infty)$,
with initial conditions

$$v(0,x) = v_0(x) \quad v_t(0,x) = v_1(x) \quad \text{for } x \in [0, L] \quad (6.38)$$

and boundary conditions

$$v(t,0) = v(t,L) = v_{xx}(t,0) = v_{xx}(t,L) = 0 \quad \text{for } t \in [0, \infty) \qquad (6.39)$$

Let

$$\Omega = \{u \in R^m = U : u_j(t) \geq 0, \text{ for } t \geq 0\},$$

Thus we have dynamical system with positive controls.

Equation (6.37) describes the transverse motion of an elastic beam which occupies the interval $[0, L]$ in the reference and stress free state. The function $w(t, x)$ denotes the displacement from the reference state at time t and position x. In the left hand side of the Eq. (6.37) the second and the third terms represent internal structural viscous damping, and the fourth term represents the effect of axial force on the beam. The boundary conditions (6.39) corresponds to hinged ends.

Let $V = L^2[0, L]$ be a separable Hilbert space of all square integrable functions on $[0, L]$ with the standard norm and inner product [31, 57]. In order to regard the vibratory system (6.37), (6.38) and (6.39) in the general framework considered in the previous sections, let us define linear unbounded differential operator

$$A : V \supset D(A) \to V$$

by [57] we have

$$Av(x) = v_{xxxx}(x) \quad \text{for} \quad v(x) \in D(A) \qquad (6.40)$$

$$D(A) = \{v(x) \in H^4[0,L]; v(0) = v(L) = v_{xx}(0) = v_{xx}(L) = 0\}$$

where $H^4[0, L]$ denotes the fourth-order Sobolev space on $[0, L]$.

The linear unbounded operator A has the following properties [31, 57]:

1. Operator A is self-adjoint and positive-definite operator with dense domain $D(A)$ in Hilbert space V.
2. There exists a compact inverse A^{-1} and consequently, the resolvent $R(s; A)$ of A is a compact operator for all $s \in \rho(A)$.
3. Operator A has a spectral representation

$$Av = \sum_{i=1}^{i=\infty} s_i \langle v, v_i \rangle_H v_i \quad \text{for} \quad v \in D(A)$$

where $s_i > 0$ and $v_i \in D(A)$, $i = 1, 2, 3, \ldots$ are simple (multiplicities $n_i=1$) eigenvalues and corresponding eigenfunctions of A, respectively. Moreover,

6.4 Constrained Controllability of Second Order Infinite Dimensional Systems

$$s_i = \left(\frac{\pi i}{L}\right)^4, \quad v_i(x) = \left(\frac{2}{L}\right)^{\frac{1}{2}} \sin\left(\frac{\pi i x}{L}\right) \quad \text{for} \quad x \in [0, L]$$

and the set $\{v_i(x), i = 1, 2, 3, \ldots\}$ forms a complete orthonormal system in V.

4. Fractional powers A^α, $0 < \alpha \leq 1$ can be defined by

$$A^\alpha v = \sum_{i=1}^{i=\infty} s_i^\alpha \langle v, v_i \rangle_H v_i, \quad \text{for} \quad v \in D(A^\alpha), \quad \text{and} \quad 0 \leq \alpha \leq 1$$

which is also a self-adjoint and positive-definite operator with a dense domain in H. Particularly, for α $^1/_2$ we have, $A^{\frac{1}{2}} v = -v_{xx}$ with the domain

$$D\left(A^{1/2}\right) = \{v \in H^2[0, L] : v(0) = v(L) = 0\}$$

Now, we can consider the partial differential Eq. (6.37) with conditions (6.38) and (6.39) as the second order evolution equation in the Hilbert space V [57].

$$\ddot{v}(t) + 2(c_2 A + c_1 A^{1/2})\dot{v}(t) + (d_2 A + d_1 A^{1/2})v(t) = \sum_{j=1}^{j=m} b_j u_j(t) \quad (6.41)$$

where

$$v(t) = v(t, \cdot) \in V$$
$$\dot{v}(t) = v_t(t, \cdot) \in V$$
$$\ddot{v}(t) = v_{tt}(t, \cdot) \in V$$
$$b_j = b_j(\cdot) \in V$$

Let the initial conditions be of the following form

$$v(0) = v_0 \in D(A)$$
$$v'(0) = v_1 \in V$$

Then there exists unique solution of the partial differential Eq. (6.37) [31].

Theorem 6.4 Vibratory dynamical system (6.37) is Ω-approximately controllable if and only if for each $i = 1, 2, 3, \ldots$ row vectors

$$B_i = \begin{bmatrix} b_{i1}, b_{i2}, \ldots b_{ij}, \ldots, b_{im} \end{bmatrix}$$

contain at least two coefficients with different signs, where

$$\sum_{j=1}^{j=m} \left(\int_0^L \sqrt{\frac{2}{L}} b_j(x) \sin\left(\frac{\pi i x}{L}\right) dx \right)^2 \neq 0 \quad \text{for } i = 1, 2, 3, \ldots \quad j = 1, 2, 3, \ldots, m$$

(6.42)

Proof Let us observe, that dynamical system (6.37) satisfies all the assumptions of Corollary 6.2. Therefore, taking into account the analytic formula for the eigenvectors $v_i(x)$, $i = 1, 2, 3 \ldots$ and the form of the inner product in the separable Hilbert space $L^2([0, L], R)$, from relation (6.37) we directly obtain inequalities (6.42). Hence, Theorem 6.4 immediately follows.

Summarizing, the present section contains results concerning approximate controllability of second order abstract infinite dimensional dynamical systems. Using the frequency-domain method [31, 57] and the methods of functional analysis and specially theory of linear unbounded operators necessary and sufficient conditions for approximate controllability in any time interval are formulated and proved.

Finally, it should be mentioned, that using general results some special cases can be also investigated and discussed. Then, the general approximate controllability conditions are applicable to approximate controllability investigation of mechanical flexible structure vibratory dynamical system [57].

The results presented in the section are generalization to second order abstract dynamical systems with damping terms of the approximate controllability conditions given in the literature [30, 31, 53, 57].

Chapter 7
Controllability and Minimum Energy Control of Fractional Discrete-Time Systems

7.1 Introduction

The reachability, controllability and minimum energy control for many kinds of linear discrete-time systems with time-delays have been considered in [14, 15, 17, 31, 52, 71]. The reachability and controllability to zero of positive fractional linear systems have been investigated in [14, 15, 17].

The non-integer order calculus both discrete and continuous was known in the mathematical literature for many years. However, up to recently its application was exclusively in mathematics. However, recently was discovered that many properties of real systems are better described with non-integer order differential or difference equations. The Chapter presents research results concerning both linear and non-linear or semilinear non-integer systems. Mathematical fundamentals of fractional calculus are given in the papers [15–17, 20].

In this paper the minimum energy control problem will be addressed for infinite-dimensional fractional discrete-time linear systems.

This chapter is organized as follows. In Sect. 7.2 the solution of the difference state equation the infinite-dimensional fractional systems is recalled. Necessary and sufficient conditions for the exact controllability of the infinite-dimensional fractional systems are established in Sect. 7.3. The main result of the paper is presented in Sect. 7.4, in which the minimum energy control problem is formulated and solved. Finally, concluding remarks are given.

To the best knowledge of the author the minimum energy control problem for the infinite-dimensional fractional discrete-time linear systems have not been considered yet.

7.2 Fractional Discrete Systems

The set of nonnegative integers will be denoted by Z_+. Let X and U be the separable generally infinite-dimensional Hilbert spaces and $x_k \in X$, $u_k \in U$, $k \in Z_+$. In finite-dimensional case $X = R^n$ and $U = R^m$.

In this paper extending [17] definition of the fractional difference of the form

$$\left[\Delta^\alpha_{x_k}\right] = \sum_{j=0}^{k}(-1)^j \binom{\alpha}{j} x_{k-j}, \tag{7.1}$$

$$n-1 < \alpha < n \in N = \{1, 2, \ldots\}, K \in Z_+$$

will be used, where $\alpha \in R$ is the order of the fractional difference and

$$\binom{\alpha}{j} = \begin{cases} 1 & \text{for } j = 0 \\ \frac{\alpha(\alpha-1)L(\alpha-j+1)}{j!} & \text{for } j = 1, 2, \ldots \end{cases} \tag{7.2}$$

Consider the fractional discrete linear system, described by the infinite-dimensional state-space equations

$$\Delta^\alpha_{x_{k+1}} = Ax_k + Bu_k, \quad k \in Z_+ \tag{7.3}$$

where $x_k \in X$, $u_k \in U$ are the state and input and $A: X \longrightarrow X$, $B: U \longrightarrow X$ are given linear and bounded operators. In finite dimensional case A and B are $n \times n$ and $n \times m$ constant matrices, respectively.

Using definition (7.1) we may write the Eqs. (7.3) in the form

$$x_{k+1} + \sum_{j=1}^{k+1}(-1)^j \binom{\alpha}{j} x_{k-j+1} = Ax_k + Bu_k, \quad k \in Z_+ \tag{7.4}$$

Lemma 7.1 [17] *The solution of Eq. (7.4) with initial condition $x_0 \in Z$ is given by*

$$x_k = \Phi_k x_0 + \sum_{i=0}^{k-1} \Phi_{k-i-1} B u_i \tag{7.5}$$

where linear and bounded operators $\Phi_k: X \longrightarrow X$ are determined by the equation

$$\Phi_{k+1} = (A + I_n \alpha)\Phi_k + \sum_{i=2}^{k+1}(-1)^{i+1}\binom{\alpha}{i}\Phi_{k-i+1} \tag{7.6}$$

with $\Phi_0 = I$, where I is the identity operator.

7.3 Controllability Conditions

First of all, in order to define controllability concepts let us introduce the notion of reachable set in q steps for infinite-dimensional discrete-time fractional control system (7.4).

Definition 7.1 For fractional system (7.4) reachable set in q steps from $x_0 = 0$ is defined as follows

$$K_q = \{x \in X : x \text{ is a solution of Eq.}(7.4) \text{ for } k = q \text{ and} \\ \text{for sequence of controls } u_0, u_1, \ldots u_k, \ldots, u_{q-1}\} \quad (7.7)$$

Remark 7.2 It should be pointed out, that in infinite-dimensional case it is necessary to distinguish between exact and approximate controllability.

Definition 7.2 The fractional system (7.4) is exactly controllable in q-steps if

$$K_q = X \quad (7.8)$$

Definition 7.3 The fractional system (7.4) is approximately controllable in q-steps if

$$cl(K_q) = X \quad (7.9)$$

where $cl(K_q)$ means the closure of the set K_q.

Theorem 7.1 *The fractional system (7.4) is exactly controllable in q steps if and only if the image $\mathrm{Im}R_q$ of controllability operator*

$$R_q = [B_0, \Phi_1 B_1, \Phi_2 B_2, \ldots, \Phi_{q-1} B_{q-1}] \quad (7.10)$$

is the whole space X.

Proof Using (7.5) for $k = q$ and $x_0 = 0$ we obtain

$$x_f = x_q = \sum_{i=0}^{q-1} \Phi_{q-i-1} B u_i = R_q \begin{bmatrix} u_{q-1} \\ u_{q-2} \\ \vdots \\ u_0 \end{bmatrix} \quad (7.11)$$

From Definition 7.2 and (7.11) it follows that for every final state $x_f \in X$ there exists a input sequence $u_i \in U$, $i = 0, 1, \ldots, q-1$ if and only if the image of controllability operator $\mathrm{Im}R_q$ is the whole space X.

Corollary 7.1 *The fractional system (7.4) is exactly controllable in q steps if and only $R_q R_q^*$ is invertible operator, i.e., there exist linear and bounded operator $\left(R_q R_q^*\right)^{-1}$.*

Corollary 7.2 *The fractional system (7.4) is approximately controllable in q steps if and only if $cl(ImR_q)$ of controllability operator (7.10) is the whole state space X, or equivalently if and only if the reachable set in q steps K_q is dense in the Hilbert space X.*

Now let us consider finite dimensional fractional dynamical systems. Since for $X = R^n$ approximate controllability in q-steps and exact controllability in q-steps coincide, we say shortly controllability in q-steps. Therefore, taking into account Theorem 7.1 we have the following Corollary.

Corollary 7.3 *The fractional finite-dimensional system (7.4) is controllable in q steps if and only if $n \times nm$—dimensional controllability matrix*

$$R_q := [B, \Phi_1 B, \ldots, \Phi_{q-1} B] \tag{7.12}$$

has full row rank n.

Example 7.1 Consider the fractional finite dimensional system (7.4) for $0 \leq \alpha \leq 1$ with

$$A = \begin{bmatrix} -\alpha & 0 \\ 1 & 2 \end{bmatrix}, \quad B = \begin{bmatrix} 1 \\ 0 \end{bmatrix}, \quad (n = 2) \tag{7.13}$$

The fractional system is positive since

$$A + I_n \alpha = \begin{bmatrix} 0 & 0 \\ 1 & 2+\alpha \end{bmatrix} \in \mathfrak{R}_+^{2 \times 2}$$

$$A + I\alpha = \begin{bmatrix} 0 & 0 \\ 2 & 1+\alpha \end{bmatrix}$$

Using (7.6) for $k = 0$ we obtain

$$\Phi_1 = (A + I_n \alpha)\Phi_0 = \begin{bmatrix} 0 & 0 \\ 1 & 2+\alpha \end{bmatrix}$$

The reachability matrix (7.10) for $q = 2$ has the form

$$R_q = [B, \ \Phi_1 B] = \begin{bmatrix} 1 & 0 \\ 0 & 1 \end{bmatrix}$$

7.3 Controllability Conditions

It contains two linearly independent monomial columns. Therefore, the fractional system with (7.13) is reachable in two steps.

Now, let us concentrate on the relationship between controllability of standard and fractional linear discrete-time systems for finite dimensional case.

The solution to the Eq. (7.3) is given by (7.5).

Substitution of (7.6) into Eq. (7.5) yields

$$x_{i+1} = (A + I_n \alpha) x_i + \sum_{j=2}^{i+1} d_j x_{i-j+1} + B u_i, \quad i \in Z_+,$$

where

$$d_j = d_j(\alpha) = (-1)^{j+1} \binom{\alpha}{j}, \quad j = 2, 3, \ldots,$$

The solution to the Eq. (7.3) has the form

$$x_i = \Phi_i x_0 + \sum_{j=0}^{i-1} \Phi_{i-j-1} B u_j,$$

where the transition state matrix can be computed using difference equation

$$\Phi_{j+1} = \Phi_j (A + I_n \alpha) + \sum_{k=2}^{j+1} d_k \Phi_{j-k+1},$$

and

$$\Phi_0 = I_n$$

Theorem 7.2 *The fractional linear discrete-time linear system (7.3) is controllable in a given interval [0, q], if and only standard discrete-time linear system with $\alpha = 1$ is controllable in the same interval [0, q], i.e. if and only if controllability matrix has full row rank,*

$$\operatorname{rank} \begin{bmatrix} B & \vdots & AB & \vdots & A^2 B & \vdots & \ldots & \vdots & A^k B & \vdots & \ldots & \vdots & A^{q-1} B \end{bmatrix} = n$$

Proof First of all, let us observe that

$$(A + I_n \alpha)^k = A^k + k \alpha A^{k-1} + \cdots + \alpha^k I_n \quad \text{for } k = 2, 3, \ldots, q-1.$$

Hence, taking into account equality given above controllability matrix for fractional system (7.3) can be expressed as follows

$$rank[\Phi_0 B \mid \Phi_1 B \mid \Phi_2 B \mid \ldots \mid \Phi_k B \mid \ldots \mid \Phi_{q-1} B] =$$
$$rank[B \mid (A+I_n\alpha)B \mid (A+I_n\alpha)^2 + c_2 I_n]B \mid \ldots \mid (A+I_n\alpha)^{q-1} + \ldots + (\alpha^{q-1} + \ldots + c_{q-1})I_n B] =$$
$$= rank[B \mid AB \mid A^2 B \mid \ldots \mid A^k B \mid \ldots \mid A^{q-1} B] \times$$
$$\times \begin{bmatrix} I_n & \alpha I_n & (d_2 + \alpha^2)I_n & \ldots & \ldots \\ 0 & I_n & 2\alpha I_n & \ldots & \ldots \\ 0 & 0 & I_n & \ldots & \ldots \\ \vdots & \vdots & \vdots & 0 & \ldots \\ 0 & 0 & 0 & \ldots & I_n \end{bmatrix}$$

Since the matrix
$$\begin{bmatrix} I_n & \alpha I_n & (d_2 + \alpha^2)I_n & \ldots & \ldots \\ 0 & I_n & 2\alpha I_n & \ldots & \ldots \\ 0 & 0 & I_n & \ldots & \ldots \\ \vdots & \vdots & \vdots & \ddots & \ldots \\ 0 & 0 & 0 & \ldots & I_n \end{bmatrix}$$

is nonsingular for all values of α and c_k, $k = 1, 2, \ldots, q-1$, then, taking into account equality (4.3) we have

$$rank[\Phi_0 B \mid \Phi_1 B \mid \Phi_2 B \mid \ldots \mid \Phi_k B \mid \ldots \mid \Phi_{q-1} B] =$$
$$= rank[B \mid AB \mid A^2 B \mid \ldots \mid A^k B \mid \ldots \mid A^{q-1} B]$$

Hence, Theorem 3.1 follows.

7.4 Minimum Energy Control

Consider the fractional infinite-dimensional system (7.4). If the system is exactly controllable in q steps then generally there exist many different input sequences that steer the initial state of the system from $x_0 = 0$ to the final state $x_f \in X$.

Among these admissible input sequences we are looking for the sequence $u_i \in U$, $i = 0, 1, \ldots, q-1$, $i \in Z_+$ that minimizes the quadratic performance index

$$I(u) = \sum_{j=0}^{q-1} u_j^T Q u_j \qquad (7.13)$$

7.4 Minimum Energy Control

where $Q: U \longrightarrow U$ is a self-adjoint positive define operator, q is a given number of steps in which the state of the system is transferred from $x_0 = 0$ to $x_f \in X$ and $u^* \in U$ denotes adjoint element, which in finite dimensional case denotes vector transposition.

The minimum energy control problem for the infinite-dimensional fractional system (7.4) can be stated as follows. For a given linear bounded operators A, B and the order α of the fractional system (7.4), the number of steps q, final state $x_f \in X$ and the selfadjoint operator Q of the performance index (7.13), find a sequence of inputs $u_i \in U, i = 0, 1, \ldots, q-1$, that steers the state of the system from initial state $x_0 = 0$ to $x_f \in X$ and minimizes the quadratic performance index (7.13).

In order to solve the minimum energy problem we define selfadjoint operator

$$W(q, Q) = R_q \bar{Q} R_q^T \tag{7.14}$$

where R_q is controllability operator defined by (7.10) and selfadjoint operator

$$\bar{Q} : \underbrace{U \times U \times \ldots \times U}_{q-times} \to \underbrace{U \times U \times \ldots \times U}_{q-times}$$

is defined as follows

$$\bar{Q} = blockdiag \lfloor Q^{-1}, Q^{-1}, \ldots, Q^{-1} \rfloor \tag{7.15}$$

From (7.15) it follows that operator $W(q, Q)$ is invertible if and only if $R_q R_q^*$ is invertible operator, i.e., there exist linear and bounded operator $\left(R_q R_q^*\right)^{-1}$ and therefore, fractional system (7.4) is exactly controllable in q steps.

If the condition of Theorem 7.1 is met then the system is exactly controllable in q steps. In this case we may define for a given $x_f \in X$ the following sequence of inputs

$$\widehat{u}_{0q} = \begin{bmatrix} \widehat{u}_{q-1} \\ \widehat{u}_{q-2} \\ \vdots \\ \widehat{u}_0 \end{bmatrix} = \bar{Q} R_q^T W^{-1}(q, Q) x_f \tag{7.16}$$

Theorem 7.3 *Let the fractional system (7.4) be exactly controllable in q steps. Moreover let $\bar{u}_i \in U, i = 0, 1, \ldots, q-1$ be a sequence of inputs that steers the state of the system from $x_0 = 0$ to $x_f \in X$. Then the sequence of inputs $\widehat{u}_i \in U, i = 0, 1, \ldots, q-1$ defined by (7.16) also steers the state of the system from $x_0 = 0$ to $x_f \in X$ and minimizes the performance index (7.13), i.e.,*

$$I(\widehat{u}) \leq I(\bar{u}) \tag{7.17}$$

The minimal value of performance index (7.13) for the minimum energy control (7.16) is given by

$$I(\hat{u}) = x_f^T W^{-1}(q, Q) x_f \qquad (7.18)$$

Proof If the fractional system (7.4) is exactly controllable in q steps, then for $x_f \in X$ we shall show that the sequence of controls given by equality (7.16) steers the state of the system (7.4) from initial state $x_0 = 0$ to final state $x_f \in X$.

Using equality (7.5) for $k = q$, $x_0 = 0$ and (7.12), (7.16) we obtain

$$x_q = R_q \hat{u}_{0q} = R_q \bar{Q} R_q^T W^{-1}(q, Q) x_f = x_f \qquad (7.19)$$

since

$$R_q \bar{Q} R_q^T W^{-1}(q, Q) = I.$$

The both sequences of inputs $\bar{u}_{0q}$ and $\hat{u}_{0q}$ steer the state of the system from $x_0 = 0$ to the same final state x_f. Hence $x_f = R_q \hat{u}_{0q} = R_q \bar{u}_{0q}$ and

$$R_q \left[\hat{u}_{0q} - \bar{u}_{0q} \right] = 0 \qquad (7.20)$$

where

Using (20) we shall show that

$$\left[\hat{u}_{0q} - \bar{u}_{0q} \right]^T \hat{Q} \hat{u}_{0q} = 0 \qquad (7.21)$$

where

$$\hat{Q} = \text{block diag}\, [Q, \ldots, Q].$$

Therefore, (7.21) yields

$$\left(\hat{u}_{0q} - \bar{u}_{oq} \right)^T R_q^T = 0 \qquad (7.22)$$

Multiplying the equality from the right by $W^{-1}(q, Q) x_f$ we obtain

$$\left(\hat{u}_{0q} - \bar{u}_{0q} \right)^T R_q^T W^{-1}(q, Q) x_f = 0 \qquad (7.23)$$

Using (7.16) and (7.23) we obtain (7.24) since

$$\begin{aligned} \left(\hat{u}_{0q} - \bar{u}_{0q} \right)^T \hat{Q} \hat{u}_{0q} &= \left(\hat{u}_{0q} - \bar{u}_{0q} \right)^T \hat{Q} \bar{Q} R_q^T W^{-1}(q, Q) x_f \\ &= \left(\hat{u}_{0q} - \bar{u}_{0q} \right)^T R_q^T W^{-1}(q, Q) x_f = 0 \end{aligned} \qquad (7.24)$$

and $\hat{Q} \bar{Q} = I$

7.4 Minimum Energy Control

Using (7.24) it is easy to verify that

$$\bar{u}_{0q}^T \hat{Q} \bar{u}_{0q} = \hat{u}_{0q}^T \hat{Q} \hat{u}_{0q} + \left[\bar{u}_{0q} - \hat{u}_{0q}\right]^T \hat{Q} \left[\bar{u}_{0q} - \hat{u}_{0q}\right] \quad (7.25)$$

From (7.25) it follows that the inequality (7.21) holds, since

$$\left[\bar{u}_{0q} - \hat{u}_{0q}\right]^T \hat{Q} \left[\bar{u}_{0q} - \hat{u}_{0q}\right] \geq 0$$

In order to find the minimal value of the performance index we substitute (7.16) into (7.15) and we use (7.16). Then we obtain

$$I(\hat{u}) = \hat{u}_{0q}^T \hat{Q} \hat{u}_{0q} = \left(\bar{Q}R_q^T W^{-1}(q,Q) x_f\right)^T \hat{Q} \left(\bar{Q}R_q^T W^{-1}(q,Q) x_f\right)$$
$$= x_f^T W^{-1}(q,Q) R_q \bar{Q} R_q^T W^{-1}(q,Q) x_f = x_f^T W^{-1}(q,Q) x_f$$

since

$$\hat{Q}\bar{Q} = I \text{ and } W^{-1}(q,Q) R_q \bar{Q} R_q^T = I$$

Example 7.2 Given the positive fractional system (7.4) for $0 < \alpha < 1$ with (7.13). Find an optimal sequence of inputs that steers the state of the system from $x_0 = 0$ to $x_f = \begin{bmatrix} 1 \\ 1 \end{bmatrix}$ in two steps ($q = 2$) and minimizes the performance index (7.15) for $Q = [4]$

In Example 7.1 it was shown that the system is reachable in two steps. It is easy to see that the conditions of Theorem 7.2 are met. Moreover, in this case

$$R_2 = [B, \Phi_1 B] = \begin{bmatrix} 1 & 0 \\ 0 & 1 \end{bmatrix}$$

$$R_2 = [B_0, B_1, \Phi_1 B_0] = \begin{bmatrix} 0 & 2 & 0 \\ 1 & 0 & 1+\alpha \end{bmatrix}$$

and

$$Q = \text{diag}\left[Q^{-1} Q^{-1}\right] = \frac{1}{2}\begin{bmatrix} 1 & 0 \\ 0 & 1 \end{bmatrix}$$

$$Q = diag[Q^{-1}, Q^{-1}] = \frac{1}{4}\begin{bmatrix} 1 & 0 \\ 0 & 1 \end{bmatrix}$$

Using (7.16) we obtain

$$W = R_q \bar{Q} R_q^T = \bar{Q} = \frac{1}{2} \begin{bmatrix} 1 & 0 \\ 0 & 1 \end{bmatrix}$$

$$W(q, Q) = W(2, 4) = R_2 \bar{Q} R_2^T = \begin{bmatrix} 2 & 0 \\ 0 & 2 \end{bmatrix}$$

Using (7.18) we obtain

$$\hat{u}_{02} = \begin{bmatrix} \hat{u}_1 \\ 0 \end{bmatrix} = \bar{Q} R_2^T W^{-1} x_f = \begin{bmatrix} 1 \\ 1 \end{bmatrix} \qquad (7.26)$$

It is easy to verify that the sequence (7.26) steers the state of the system in two steps from $x_0 = 0$ to $x_f = \begin{bmatrix} 1 & 1 \end{bmatrix}^T$.

The minimal value of the performance index in this case is equal to 4.

$$I(\hat{u}) = x_f^T W^{-1} x_f = \begin{bmatrix} 1 & 1 \end{bmatrix} \begin{bmatrix} 2 & 0 \\ 0 & 2 \end{bmatrix} \begin{bmatrix} 1 \\ 1 \end{bmatrix} = 4$$

The minimum energy control problem of infinite-dimensional fractional discrete linear systems has been addressed. Necessary and sufficient conditions for the exact controllability in q steps of the systems have been established.

Under assumption on exact controllability in q steps solvability of the minimum energy control of the infinite-dimensional fractional discrete-time linear systems have been given and a procedure for computation of the optimal sequence of inputs minimizing the quadratic performance index has been proposed.

Finally, it should be mentioned, that the considerations can be extended for infinite-dimensional fractional discrete-time linear systems with delays both in control and state variables and for infinite-dimensional fractional continuous-time linear systems with constant parameters.

Chapter 8
Controllability and Minimum Energy Control of Fractional Discrete-Time Systems with Delay

8.1 Introduction

The present Chapter is organized as follows. In Sect. 8.2 the solution of the difference state equation finite-dimensional fractional systems with delay is recalled. Necessary and sufficient conditions for controllability of the are established in Sect. 8.3. The main result of the Chapter is presented in Sect. 8.4, in which the minimum energy control problem is formulated and solved. Finally, concluding remarks are given at the end of Chapter.

To the best knowledge of the author controllability and minimum energy control problem for fractional discrete-time linear systems with delay in control have not been considered yet.

8.2 Fractional Delayed Systems

Similarly as in Chap. 7 the set of nonnegative integers will be denoted by Z_+. Let $x_k \in R^n$, $u_k \in R^m$, $k \in Z_+$.

In this chapter extending definition of the fractional difference of the form

$$\Delta^\alpha_{x_k} = \sum_{j=0}^{k} (-1)^j \binom{\alpha}{j} x_{k-j},$$

$$n - 1 < \alpha < n \in N = \{1, 2, \ldots\}, \; k \in Z_+$$

will be used, where $\alpha \in R$ is the order of the fractional difference and

$$\binom{\alpha}{j} = \begin{cases} 1 & \text{for } j = 0 \\ \frac{\alpha(\alpha-1)\ldots(\alpha-j+1)}{j!} & \text{for } j = 1, 2, \ldots \end{cases}$$

Consider the fractional discrete linear system, described by the difference state-space equations with delayed control

$$\Delta^\alpha_{x_{k+1}} = Ax_k + B_0 u_k + B_1 u_{k-1} \qquad (8.1)$$

where $x_k \in R^n$, $u_k \in R^m$ are the state and input and A and B are $n \times n$ and $n \times m$ constant matrices, respectively.

Using fractional difference definition we may write the Eq. (8.1) in the form

$$x_{k+1} + \sum_{j=1}^{j=k+1} (-1)^j \binom{\alpha}{j} x_{k-j+1} = Ax_k + B_0 u_k + B_1 u_{k-1} \qquad (8.2)$$

Lemma 8.1 *Reference* [17]. *The solution of Eq. (8.4) with initial condition $x_0 \in R^n$ is given by*

$$\begin{aligned} x_k &= \Phi_k x_0 + \sum_{i=0}^{i=k-1} (\Phi_{k-i-1}(B_0 u_i + B_1 u_{i-1})) \\ &= \Phi_k x_0 + \Phi_{k-1} B_1 u_{-1} + \sum_{i=0}^{i=k-1} (\Phi_{k-i-1} B_0 + \Phi_{k-i-2} B_1) u_i \end{aligned} \qquad (8.3)$$

where $n \times n$ dimensional matrices Φ_k, $k = 0,1,2,\ldots$ are determined by the equation

$$\Phi_{k+1} = (A + I_n \alpha)\Phi_k + \sum_{i=2}^{k+1} (-1)^{i+1} \binom{\alpha}{i} \Phi_{k-i+1} \qquad (8.4)$$

with $\Phi_0 = I$, where I is $n \times n$ dimensional identity matrix and $\Phi_k = 0$ for $k < 0$.

8.3 Controllability Conditions

First of all, in order to define controllability concepts let us introduce the notion of reachable set in q steps for infinite-dimensional discrete-time fractional control system (8.2).

Definition 8.1 For fractional system (8.2) controllable set in q steps from $x_0 = 0$ is defined as follows

$$\begin{aligned} K_q = \{x \in X : x \text{ is a solution of Eq. (8.4)} \\ \text{for } k = q \text{ and for sequence of controls } u_0, u_1, \ldots u_k, \ldots, u_{q-1}\} \end{aligned} \qquad (8.5)$$

8.3 Controllability Conditions

Definition 8.2 The fractional system (8.2) is controllable in q-steps if

$$K_q = R^n \qquad (8.6)$$

Let us introduce the $n \times nm$ dimensional controllability matrix

$$R_q = [B_0, (\Phi_1 B_0 + B_1), (\Phi_2 B_0 + \Phi_1 B_1), \\ \ldots, (\Phi_{k-i-1} B_0 + \Phi_{k-i-2} B_1), \ldots, (\Phi_{q-1} B_0 + \Phi_{q-2} B_1)] \qquad (8.7)$$

Theorem 8.1 *The fractional system with delay in control (8.2) is controllable in q steps if and only if*

$$\text{rank } R_q = n \qquad (8.8)$$

Proof Using (8.3) for $k = q$, $x_0 = 0$, $u_{-1} = 0$ we obtain

$$x_f = x_q = \sum_{i=0}^{i=k-1} (\Phi_{k-i-1} B_0 + \Phi_{k-i-2} B_1) u_i = R_q \begin{bmatrix} u_{q-1} \\ u_{q-2} \\ \vdots \\ u_i \\ \vdots \\ u_0 \end{bmatrix} \qquad (8.9)$$

From Definition 8.2 and (8.9) it follows that for every final state $x_f \in R^n$ there exists a input sequence $u_i \in R^m$, $i = 0, 1, \ldots, q-1$ if and only if the image of controllability matrix $Im R_q$ is the whole space R^n, so controllability matrix R_q has full row rank n.

Corollary 8.1 *The fractional system (8.2) is controllable in q steps if and only $n \times n$ dimensional constant matrix $R_q R_q^T$ is invertible, i.e., there exist inverse matrix $\left(R_q R_q^T\right)^{-1}$.*

8.4 Minimum Energy Control

Consider the fractional system with delayed control (8.2). If the system is controllable in q steps then generally there exist many different input sequences that steer the initial state of the system from $x_0 = 0$, $u_{-1} = 0$ to the final state $x_f \in R^n$.

Among these input sequences we are looking for the sequence $u_i \in U$, $i = 0, 1, \ldots, q-1$, $i \in Z_+$ that minimizes the quadratic performance index

$$I(u) = \sum_{j=0}^{q-1} u_j^T Q u_j \qquad (8.10)$$

where Q is a $m \times m$ dimensional positive define matrix, q is a given number of steps in which the state of the system is transferred from $x_0 = 0$, $u_{-1} = 0$, to $x_f \in X$ and $u^T \in R^m$ denotes vector transposition.

The minimum energy control problem for fractional system with delayed control (8.2) can be stated as follows. For a given matrices A, B and the order α of the fractional system (8.2), the number of steps q, final state $x_f \in R^n$ and the positive definite matrix Q of the performance index (8.10), find a sequence of inputs $u_i \in R^m$, $i = 0, 1, \ldots, q-1$, that steers the state of the system from initial state $x_0 = 0$, $u_{-1} = 0$, to final state $x_f \in R^n$ and minimizes the quadratic performance index given by equality (8.10).

In order to solve the minimum energy problem we define $n \times n$ dimensional symmetric matrix

$$W(q, Q) = R_q \bar{Q} R_q^T \qquad (8.11)$$

where R_q is controllability matrix defined by (8.8) and $qm \times qm$ dimensional matrix $\bar{Q}$ is defined as follows

$$\bar{Q} = blockdiag\left[Q^{-1}, Q^{-1}, \ldots, Q^{-1}\right] \qquad (8.12)$$

From (8.12) it follows that matrix $W(q, Q)$ is invertible if and only if $R_q R_q^T$ matrix is invertible and therefore, fractional system (8.2) is controllable in q steps.

If the condition of Theorem 8.1 is met then the system is exactly controllable in q steps. In this case we may define for a given $x_f \in R^n$ the following sequence of inputs

$$\widehat{u}_{0q} = \begin{bmatrix} \widehat{u}_{q-1} \\ \widehat{u}_{q-2} \\ \vdots \\ \widehat{u}_0 \end{bmatrix} = \bar{Q} R_q^T W^{-1}(q, Q) x_f \qquad (8.13)$$

Theorem 8.2 *Let the fractional system (8.2) be controllable in q steps. Moreover let $\bar{u}_i \in R^m$, $i = 0, 1, \ldots, q-1$ be a sequence of inputs that steers the state of the system from $x_0 = 0$, $u_{-1} = 0$ to $x_f \in R^n$. Then the sequence of inputs $\widehat{u}_i \in R^m$, $i = 0, 1, \ldots, q-1$ defined by (8.13) also steers the state of the system from $x_0 = 0$, $u_{-1} = 0$ to $x_f \in R^n$ and minimizes the performance index (8.10), i.e.,*

$$I(\widehat{u}) \leq I(\bar{u}) \qquad (8.14)$$

The minimal value of performance index (8.10) for the minimum energy control (8.13) is given by

$$I(\widehat{u}) = x_f^T W^{-1}(q, Q) x_f \qquad (8.15)$$

8.4 Minimum Energy Control

Proof If the fractional system (8.2) is controllable in q steps, then for $x_f \in R^n$ we shall show that the sequence of controls given by equality (8.13) steers the state of the system (8.2) from initial state $x_0 = 0$, $u_{-1} = 0$ to final state $x_f \in R^n$.

Using equality (8.3) for $k = q$, $x_0 = 0$, $u_{-1} = 0$ and (8.9), (8.13) we obtain

$$x_q = R_q \hat{u}_{0q} = R_q \bar{Q} R_q^T W^{-1}(q, Q) x_f = x_f \qquad (8.16)$$

since

$$R_q \bar{Q} R_q^T W^{-1}(q, Q) = I.$$

The both sequences of inputs $\bar{u}_{0q}$ and $\hat{u}_{0q}$ steer the state of the system from $x_0 = 0$, $u_{-1} = 0$ to the same final state x_f.

Hence

$$x_f = R_q \hat{u}_{0q} = R_q \bar{u}_{0q}$$

and

$$R_q [\hat{u}_{0q} - \bar{u}_{0q}] = 0 \qquad (8.17)$$

Using (8.17) we shall show that

$$[\hat{u}_{0q} - \bar{u}_{0q}]^T \hat{Q} \hat{u}_{0q} = 0 \qquad (8.18)$$

where $\hat{Q} = \text{block diag }[Q, \ldots, Q]$.

Therefore, (8.18) yields

$$\left(\hat{u}_{0q} - \bar{u}_{oq}\right)^T R_q^T = 0 \qquad (8.19)$$

Multiplying the above equality by $W^{-1}(q, Q) x_f$ we obtain

$$\left(\hat{u}_{0q} - \bar{u}_{oq}\right)^T R_q^T W^{-1}(q, Q) x_f = 0 \qquad (8.20)$$

Using (8.13) and (8.20) and since

$$\left(\hat{u}_{0q} - \bar{u}_{0q}\right)^T \hat{Q} \hat{u}_{0q}$$
$$= \left(\hat{u}_{0q} - \bar{u}_{0q}\right)^T \hat{Q} \bar{Q} R_q^T W^{-1}(q, Q) x_f$$
$$= \left(\hat{u}_{0q} - \bar{u}_{0q}\right)^T R_q^T W^{-1}(q, Q) x_f = 0$$

and
$$\hat{Q}\bar{Q} = I$$

it is easy to verify that

$$\bar{u}_{0q}^T \bar{Q} \hat{u}_{0q} = \hat{u}_{0q}^T \hat{Q} \hat{u}_{0q} + [\bar{u}_{0q} - \hat{u}_{0q}]^T \hat{Q} [\bar{u}_{0q} - \hat{u}_{0q}] \qquad (8.21)$$

From (8.21) it follows that the inequality (8.14) holds, since

$$[\bar{u}_{0q} - \hat{u}_{0q}]^T \hat{Q} [\bar{u}_{0q} - \hat{u}_{0q}] \geq 0$$

In order to find the minimal value of the performance index we substitute (8.13) into (8.12) and we use (8.13). Then we obtain

$$\begin{aligned} I(\hat{u}) &= \hat{u}_{0q}^T \hat{Q} \hat{u}_{0q} \\ &= \left(\bar{Q} R_q^T W^{-1}(q, Q) x_f \right)^T \hat{Q} \left(\bar{Q} R_q^T W^{-1}(q, Q) x_f \right) \\ &= x_f^T W^{-1}(q, Q) R_q \bar{Q} R_q^T W^{-1}(q, Q) x_f = x_f^T W^{-1}(q, Q) x_f \end{aligned} \qquad (8.22)$$

since $\hat{Q}\bar{Q} = I$ and $W^{-1}(q, Q) R_q \bar{Q} R_q^T = I$

Hence, Theorem follows.

Example 8.1 Consider the fractional system with delayed control of the form (8.2) for $0 \leq \alpha \leq 1$ with the following matrices

$$A = \begin{bmatrix} -\alpha & 0 \\ 2 & 1 \end{bmatrix} \quad B_0 = \begin{bmatrix} 0 \\ 1 \end{bmatrix} \quad B_1 = \begin{bmatrix} 1 \\ 0 \end{bmatrix} \qquad (8.23)$$

Hence

$$A + I\alpha = \begin{bmatrix} 0 & 0 \\ 2 & 1+\alpha \end{bmatrix}$$

Using (8.4) for $k = 0$ we obtain

$$\Phi_1 = (A + I\alpha)\Phi_0 = \begin{bmatrix} 0 & 0 \\ 2 & 1+\alpha \end{bmatrix}$$

The reachability matrix (8.8) for $q = 2$ has the form

$$R_2 = [B_0, (\Phi_1 B_0 + B_1)] = \begin{bmatrix} 0 & 1 \\ 1 & 1+\alpha \end{bmatrix}$$

8.4 Minimum Energy Control

Therefore, *rank* $R_2 = 2 = n$ and the fractional system with delayed control (8.23) is controllable in two steps.

Since it was shown that the system is controllable in two steps then it is easy to see that the conditions of Theorem 8.2 are met.

Therefore, find an optimal sequence of inputs that steers the state of the system from $x_0 = 0$, $u_{-1} = 0$, to $x_f = \begin{bmatrix} 1 \\ 2 \end{bmatrix}$ in two steps ($q = 2$) and minimizes the performance index (8.10) for $Q = [4]$.

First, by

$$\bar{Q} = diag\left[Q^{-1}, Q^{-1}\right] = \frac{1}{4}\begin{bmatrix} 1 & 0 \\ 0 & 1 \end{bmatrix}$$

Using (8.13) we obtain

$$W(q, Q) = W(2, 4) = R_2 \bar{Q} R_2^T$$

$$= \frac{1}{4}\begin{bmatrix} 0 & 1 \\ 1 & 1+\alpha \end{bmatrix}\begin{bmatrix} 1 & 0 \\ 0 & 1 \end{bmatrix}\begin{bmatrix} 0 & 1 \\ 1 & 1+\alpha \end{bmatrix}$$

$$= \frac{1}{4}\begin{bmatrix} 1 & 1+\alpha \\ 1+\alpha & 2+2\alpha+\alpha^2 \end{bmatrix}$$

Next, taking into account equality (8.15) we obtain

$$\hat{u}_{02} = \begin{bmatrix} \hat{u}_1 \\ \hat{u}_0 \end{bmatrix} = \bar{Q}R_2^T W^{-1}(2, 4) x_f$$

$$= \begin{bmatrix} 0 & 1 \\ 1 & 1+\alpha \end{bmatrix}\begin{bmatrix} 2+2\alpha+\alpha^2 & -(1+\alpha) \\ -(1+\alpha) & 1 \end{bmatrix}\begin{bmatrix} 1 \\ 2 \end{bmatrix}$$

$$= \begin{bmatrix} -(1+\alpha) & 1 \\ 2+2\alpha+\alpha^2 & 0 \end{bmatrix}\begin{bmatrix} 1 \\ 2 \end{bmatrix} = \begin{bmatrix} 1-\alpha \\ 2+2\alpha+\alpha^2 \end{bmatrix}$$

It is easy to verify that the above sequence steers the state of the system in two steps from $x_0 = 0$, $u_{-1} = 0$, to $x_f = \begin{bmatrix} 1 \\ 2 \end{bmatrix}$.

The minimal value of the performance index (8.10) in this case is equal to

$$I(\hat{u}) = x_f^T W^{-1}(q, Q) x_f$$

$$= \frac{1}{4}[1 \; 2]\begin{bmatrix} 2+2\alpha+\alpha^2 & -(1+\alpha) \\ -(1+\alpha) & 1 \end{bmatrix}\begin{bmatrix} 1 \\ 2 \end{bmatrix}$$

$$= \frac{1}{4}(\alpha^2 - 2\alpha + 2)$$

The minimum energy control problem of fractional discrete linear systems with delayed controls has been addressed. Necessary and sufficient conditions for the exact controllability in q steps of the systems have been established.

Under assumption on controllability in q steps solvability of the minimum energy control of fractional discrete-time linear systems with delayed controls have been given and a procedure for computation of the optimal sequence of inputs minimizing the quadratic performance index has been proposed.

Finally, it should be mentioned, that the considerations can be extended for fractional discrete-time linear systems with delays in state variables and for infinite-dimensional fractional discrete-time linear systems with constant parameters.

Chapter 9
Controllability of Fractional Discrete-Time Semilinear Systems

9.1 Introduction

Controllability problems studied in this Chapter concern semilinear fractional discrete-time control systems. More precisely, in the present paper unconstrained local controllability problem of finite-dimensional fractional discrete-time semilinear systems is addressed.

Using general formula of solution of difference state equation, sufficient condition for local controllability in a given number of steps is formulated and proved. The present paper extends for semilinear discrete-time fractional control systems with constant coefficients controllability results given in [15, 17, 71] for linear fractional systems.

The Chapter is organized as follows. In Sect. 8.2 using results presented in [17], general solution of the difference state equation for finite-dimensional fractional linear systems is recalled. Sufficient condition for local unconstrained controllability of the semilinear fractional discrete-time control system with constant parameters is established in Sect. 8.3. Section 8.4 contains simple numerical example, which illustrates theoretical considerations. Finally, concluding remarks and propositions for future works are given at the end of the Chapter.

9.2 Fractional Semilinear Systems

Let us consider the fractional discrete-time linear system, described by the semilinear difference state-space equation

$$\Delta^{\alpha}_{x_{k+1}} = Ax_k + Bu_k + f(x_k, u_k) \tag{9.1}$$

where $x_k \in R^n$, $u_k \in R^m$ are the state and input and A and B are $n \times n$ and $n \times m$ constant matrices, $f: R^n \times R^m \to R^n$ is nonlinear function differentiable near zero in the space $R^n \times R^m$ and such that $f(0, 0) = 0$.

Let us observe, that semilinear discrete-time control system is described by the difference state equation, which contains both pure linear and pure nonlinear parts in the right hand side of the state equation.

Using definition of fractional difference given in previous Chapters we may write semilinear difference Eq. (9.1) in the equivalent form.

$$x_{k+1} + \sum_{j=1}^{j=k+1} (-1)^j \binom{\alpha}{j} x_{k-j+1} = Ax_k + Bu_k + f(x_k, u_k)$$

Next, using standard linearization method it is possible to find the associated linear difference state equation

$$x_{k+1} + \sum_{j=1}^{j=k+1} (-1)^j \binom{\alpha}{j} x_{k-j+1} = Ax_k + Bu_k + Fx_k + Gu_k$$

where $n \times n$ dimensional matrix

$$F = \frac{d}{dx} f(x, u) \bigg|_{\substack{x=0, \\ u=0}}$$

and $n \times m$ dimensional matrix

$$G = \frac{d}{du} f(x, u) \bigg|_{\substack{x=0, \\ u=0}},$$

Moreover, for simplicity of notation let us denote

$$A + F = C \text{ and } D = B + G.$$

Thus we have

$$x_{k+1} + \sum_{j=1}^{j=k+1} (-1)^j \binom{\alpha}{j} x_{k-j+1} = Cx_k + Du_k \qquad (9.2)$$

9.2 Fractional Semilinear Systems

Lemma 9.1 Kaczorek [15, 17]. *The solution of linear difference Eq. (9.2) with initial condition $x_0 \in R^n$ is given by*

$$x_k = \Phi_k x_0 + \sum_{i=0}^{i=k-1} (\Phi_{k-i-1} D u_i) \tag{9.3}$$

where $n \times n$ dimensional state transition matrices Φ_k, $k = 0, 1, 2,\ldots$ are determined by the recurrent formula

$$\Phi_{k+1} = (C + I_n \alpha)\Phi_k + \sum_{i=2}^{i=k+1} (-1)^{j+1} \binom{\alpha}{i} \Phi_{k-i+1} \tag{9.4}$$

with $\Phi_0 = I_n$, where I_n is $n \times n$ dimensional identity matrix and by assumption matrices $\Phi_k = 0$ for $k < 0$.

Moreover, it should be pointed out, that the matrices Φ_k, $k = 0, 1, 2,\ldots$ defined above are extensions for fractional linear discrete-time control systems, the well known state transition matrices [31] for standard linear discrete-time control systems.

9.3 Controllability Conditions

First of all, in order to define global and local controllability concepts for semilinear and linear finite-dimensional discrete-time control systems let us introduce the notion of reachable set or in other words attainable set in q steps [15, 17, 31, 32].

Definition 9.1 For fractional semilinear system (9.1) or linear system (9.2) reachable set in q steps from initial condition $x_0 = 0$ is defined as follows: 0020 (9.1), (9.2)

$$K_q = \left\{ \begin{array}{l} x(q) \in R^n : x(q) \text{ is a solution of semilinear system or} \\ \text{linear system in step } q \text{ for sequence of admissible controls } u_0, u_1, \ldots u_k, \ldots, u_{q-1} \end{array} \right\} \tag{9.5}$$

Definition 9.2 The fractional semilinear discrete-time control system (9.1) is locally controllable in q-steps if there exists a neighborhood of zero $N \subset R^n$, such that

$$K_q = N \tag{9.6}$$

Definition 9.3 The fractional linear discrete-time linear control system (9.2) is globally controllable in q-steps if

$$K_q = N \tag{9.7}$$

For linear control system (9.2) let us introduce the $n \times qm$ dimensional controllability matrix

$$H_q = \left[D, (\Phi_1 D), (\Phi_2 D), \ldots, (\Phi_i D), \ldots, (\Phi_{q-1} D)\right]. \tag{9.8}$$

In order to prove sufficient condition for local controllability of semilinear discrete-time fractional control systems (9.1), we shall use certain result taken directly from nonlinear functional analysis. This result concerns so called nonlinear covering operators.

Lemma 9.2 Robinson [68] let $W: Z \to Y$ be a nonlinear operator from a Banach space Z into a Banach space Y and $W(0) = 0$. Moreover, it is assumed, that operator W has the Frechet derivative $dW(0): Z \to Y$, whose image coincides with the whole space Y. Then the image of the operator W will contain a neighborhood of the point $W(0) \in Y$.

Now, we are in the position to formulate and prove the main result on the local unconstrained controllability in the interval $[0, q]$ for the nonlinear discrete-time system (9.1). This result is known for semilinear or nonlinear continuous-time control system and is given in [33, 34], as a sufficient condition for local controllability.

Theorem 9.1 Semilinear discrete-time control system (9.1) is locally controllable in q steps if the associated linear discrete-time control system (9.2) is globally controllable in q-steps.

Proof Proof of the Theorem 9.1 is based on Lemma 9.1 and Lemma 9.2. Let the nonlinear operator W transform the space of admissible control sequence $\{u(i): 0 \leq i \leq q\}$ into the space of solutions at the step q for the semilinear discrete-time fractional control system (9.1).

More precisely, the nonlinear operator

$$W : R^m \times R^m \times \cdots \times R^m \to R^n$$

associated with semilinear control system (9.1) is defined as follows [31, 33],

$$W\{u(0), u(1), u(2), \ldots, u(i), \ldots, u(q-1)\} = x_{sem}(q)$$

where $x_{sem}(q)$ is the solution at the step q of the semilinear discrete-time fractional control system (9.1) corresponding to an admissible controls sequence $u_q = \{u(i): 0 \leq i < q\}$.

Therefore, for zero initial condition Frechet derivative at point zero of the nonlinear operator W denoted as $dW(0)$ is a linear bounded operator defined by the following formula

$$dW(0)\{u(0), u(1), u(2), \ldots, u(i), \ldots, u(q-1)\} = x_{lin}(q)$$

where $x_{lin}(q)$ is the solution at the step q of the linear system (9.2) corresponding to an admissible controls sequence $u_q = \{u(i): 0 \leq i < q\}$ for zero initial condition.

9.3 Controllability Conditions

Since from the assumption nonlinear function $f(0, 0) = 0$, then for zero initial condition the nonlinear operator W transforms zero in the space of admissible controls into zero in the state space, i.e., $W(0) = 0$.

Moreover, let us observe, that if the associated linear discrete-time fractional control system (9.2) is globally controllable in the interval $[0, q]$, then by Definition 9.1 the image of the Frechet derivative $dW(0)$ covers whole state space R^n.

Therefore, by the result stated at the beginning of the proof, the nonlinear operator W covers some neighborhood of zero in the state space R^n. Hence, by Definition 9.2 semilinear discrete-time fractional control system (9.1) is locally controllable in the interval $[0, q]$. This completes the proof.

Now, for the convenience, let us recall some well known [15, 17, 31, 33] facts from the controllability theory of linear finite-dimensional discrete-time fractional control systems.

Theorem 9.2 Klamka [50] *the fractional discrete-time linear system (9.2) is globally controllable in q steps if and only if*

$$\text{rank } H_q = n \tag{9.9}$$

Taking into account the form of controllability matrix, from Theorem 9.2 *immediately follows the simple Corollary.*

Corollary 9.1 Klamka [50] *the fractional linear control system (9.2) is controllable in q steps if and only if $n \times n$ dimensional constant matrix $H_q H_q^T$ is invertible, i.e. there exists the inverse matrix $\left(H_q H_q^T\right)^{-1}$.*

Corollary 9.2 *The fractional semilinear control system (9.1) is controllable in q steps if equality (9.9) holds or equivalently if $n \times n$ dimensional constant matrix $H_q H_q^T$ is invertible, i.e. there exists the inverse matrix $\left(H_q H_q^T\right)^{-1}$.*

Example 9.1 Let us consider the semilinear fractional discrete-time control system with constant coefficients of the form (9.1) for $0 \leq \alpha \leq 1$ with the following matrices and vectors in the difference state equation.

$$A = \begin{bmatrix} 1 & 0 \\ 0 & 1 \end{bmatrix},$$

$$B = \begin{bmatrix} 0 \\ 1 \end{bmatrix}, \tag{9.10}$$

$$f(x, u) = f(x_1, x_2, u) = \begin{bmatrix} e^u - 1 \\ 2 \sin x_1 \end{bmatrix}$$

Hence we have

$$f(0,0,0) = \begin{bmatrix} 0 \\ 0 \end{bmatrix}$$

$$F = \frac{d}{dx}f(x_1,x_2,u)\bigg|_{\substack{x=0 \\ u=0}} = \begin{bmatrix} 0 & 0 \\ 2 & 0 \end{bmatrix}$$

$$G = \frac{d}{du}f(x_1,x_2,u)\bigg|_{\substack{x=0 \\ u=0}} = \begin{bmatrix} 1 \\ 0 \end{bmatrix}$$

Hence we have

$$C = A + F = \begin{bmatrix} 1 & 0 \\ 2 & 1 \end{bmatrix},$$

$$D = B + G = \begin{bmatrix} 1 \\ 1 \end{bmatrix}$$

Using formula (9.4) for $k = 0$ we obtain

$$\Phi_1 = (C + I\alpha)\Phi_0 = \begin{bmatrix} 1+\alpha & 0 \\ 2 & 1+\alpha \end{bmatrix}$$

Controllability matrix (10) for $q = 2$ has the form

$$H_2 = [D, (\Phi_1 D)] = \begin{bmatrix} 1 & 1+\alpha \\ 1 & 3+\alpha \end{bmatrix}$$

Therefore, since *rank* $H_2 = 2 = n$ then taking into account Theorem 9.2 the fractional associated linear discrete-time system with constant coefficients is globally controllable in two steps, hence by Theorem 9.1 the semilinear fractional discrete-time system (9.10) is locally controllable in two steps.

For comparison let us consider linear fractional discrete system (9.2) with the matrices A and B given equalities (9.10). In this case using formula (9.4) for $k = 0$ we have

$$\Phi_1 = (A + I\alpha)\Phi_0 = \begin{bmatrix} 1+\alpha & 0 \\ 0 & 1+\alpha \end{bmatrix}$$

9.3 Controllability Conditions

Controllability matrix (9.8) for $q = 2$ has the form

$$H_2 = [B, (\Phi_1 B)] = \begin{bmatrix} 0 & 0 \\ 1 & 1+\alpha \end{bmatrix}$$

Therefore, since *rank* $H_2 = 1 < n$ then taking into account Corollary 9.1 the fractional linear discrete-time system with constant coefficients is not globally controllable in two steps and consequently in any number of steps.

Example 9.2 Let us consider the following simple example, which illustrates theoretical considerations. Let us assume unconstrained admissible controls and let the semilinear fractional discrete dynamical control system defined on a given time interval [0, q], has the following form

$$\Delta_{k+1}^{1,5} x_1(k+1) = x_1(k) + x_2(k) + \exp x_1(k) - \cos x_2(k) + \sin u(k)$$
$$\Delta_{k+1}^{1,5} x_2(k+1) = -x_1(k) + x_2(k) + \cos x_1(k) + \sin x_2(k) - \exp u(k) \qquad (9.11)$$

Therefore,

$$n = 2, \ m = 1, \ x(t) = (x_1(t), x_2(t))^{tr} \in R^2, U = R, \alpha = 1, 5$$

and using the notations given in the previous sections matrices C and D and the nonlinear mapping F have the following form

$$C = \begin{bmatrix} 1 & 1 \\ -1 & 1 \end{bmatrix}$$

$$D = \begin{bmatrix} 0 \\ 1 \end{bmatrix}$$

$$f(x(k), u(k)) = \begin{bmatrix} \exp x_1(k) - \cos_2(k) + \sin u(k) \\ \cos x_1(k) + \sin x_2(k) - \exp u(k) \end{bmatrix}$$

Moreover,

$$f(0,0) = \begin{bmatrix} 0 \\ 0 \end{bmatrix}$$

$$f_x(x(k), u(k)) = \begin{bmatrix} \exp x_1(k) & \sin x_2(k) \\ -\sin x_2(k) & \cos x_2(k) \end{bmatrix}$$

$$f_x(0,0) = \begin{bmatrix} 1 & 0 \\ 0 & 1 \end{bmatrix}$$

$$f_u(x(k), u(k)) = \begin{bmatrix} \cos u(k) \\ -\exp u(k) \end{bmatrix}$$

$$f_u(0,0) = \begin{bmatrix} 1 \\ -1 \end{bmatrix}$$

$$A = C + f_x(0,0) = \begin{bmatrix} 2 & 1 \\ -1 & 2 \end{bmatrix}$$

$$B = D + f_u(0,0) = \begin{bmatrix} 0 \\ 1 \end{bmatrix} + \begin{bmatrix} 1 \\ -1 \end{bmatrix} = \begin{bmatrix} 1 \\ 0 \end{bmatrix}$$

Moreover, taking into account the equality (9.5) for $k = 0$ and $\alpha = 1,5$ we have

$$S_{k+1} = (A + \alpha I_n) S_k + \sum_{i=1}^{i=k+1} (-1)^{i+1} \binom{\alpha}{i} S_{k-i+1}$$

$$= S_1 = (A + 1,5 I_2) S_0 - \binom{1,5}{1} S_0 = (A + 1,5 I_2) I_2 - 1,5 I_2$$

Thus substituting matrices A and I_2 we obtain

$$S_1 = \begin{bmatrix} 2 & 1 \\ -1 & 2 \end{bmatrix} + 1,5 \begin{bmatrix} 1 & 0 \\ 0 & 1 \end{bmatrix} = \begin{bmatrix} 3,5 & 1 \\ -1 & 3,5 \end{bmatrix}$$

Thus for $q = 2$ steps controllability matrix M_2 has the form

$$M_2 = [B \quad S_1 B] = \begin{bmatrix} 1 & 3,5 \\ 0 & -1 \end{bmatrix}$$

Since *rank* $M_2 = 2 = n$ the linear fractional system is globally controllable in two steps and hence, the semilinear fractional discrete system (9.11) is locally controllable in two steps near origin.

In the present Chapter unconstrained local controllability problem of finite-dimensional fractional discrete-time semilinear systems has been addressed. Using linearization method and solution formula for linear difference equation sufficient condition for unconstrained local controllability in q steps of the discrete-time fractional control system has been established as rank condition of suitably defined controllability matrix.

In the proof of the main result certain theorem taken directly from nonlinear functional analysis has been used. Moreover, simple illustrative numerical example has been also presented.

9.3 Controllability Conditions

There are many possible extensions of the results given in the paper. First of all it is possible to consider semilinear infinite-dimensional fractional control systems. Moreover, it should be mentioned, that controllability considerations presented in the paper can be extended for fractional discrete-time linear systems with multiple delays both in the controls and in the state variables.

Chapter 10
Controllability of Fractional Discrete-Time Semilinear Systems with Multiple Delays in Control

10.1 Introduction

Controllability problems studied in this Chapter concern semilinear fractional discrete-time control systems with multiple delays in control. More precisely, in the present paper unconstrained local controllability problem of finite-dimensional fractional-discrete time semilinear systems with multiple delays in control is addressed.

Using general formula of solution of difference state equation, sufficient condition for local controllability in a given number of steps is formulated and proved. The present paper extends for semilinear discrete-time fractional control systems with constant coefficients controllability results given in [15, 31–33] for linear discrete or fractional systems.

The Chapter is organized as follows. In Sect. 10.2 using results presented in [31], general solution of the difference state equation for finite-dimensional fractional linear systems with delays in control is recalled. Sufficient condition for local unconstrained controllability of the semilinear fractional discrete-time control system with constant parameters is established in Sect. 10.3. Section 10.4 contains simple numerical example, which illustrates theoretical considerations. Finally, concluding remarks and propositions for future works are given at the end of the Chapter.

10.2 Fractional Semilinear Systems with Multiple Delays in Control

Let us consider the fractional discrete linear system, described by the semilinear difference state-space equation

$$\Delta^\alpha x_{k+1} = A x_k + \sum_{j=0}^{j=p} B_j u_{k-j} + f(x_k, u_k, u_{k-1}, \ldots u_{k-j}, \ldots, u_{k-p}) \qquad (10.1)$$

where $x_k \in R^n, u_k \in R^m$ are the state and input,

$A, B_j, j = 0, 1, 2, \ldots, p$ are $n \times n$ and $n \times m$ constant matrices,

$f : R^n \times R^m \times \cdots \times R^m \to R^n$ is nonlinear function differentiable near zero in the space $R^n \times R^m \times \cdots \times R^m$ and such that $f(0, 0, \ldots, 0) = 0$.

Let us observe, that semilinear discrete-time control system is described by the difference state equation, which contains both pure linear and pure nonlinear parts in the right hand side of the state equation.

Using definition of fractional difference we may write semilinear difference Eq. (10.1) in the equivalent form

$$x_{k+1} + \sum_{j=1}^{j=k+1} (-1)^j \binom{\alpha}{j} x_{k-j+1} = A x_k + \sum_{j=0}^{j=p} B_j u_{k-j} + f(x_k, u_k, u_{k-1}, \ldots, u_{k-j}, \ldots, u_{k-j})$$

Next, using standard linearization method [33] it is possible to find the associated linear difference state equation

$$x_{k+1} + \sum_{j=1}^{j=k+1} (-1)^j \binom{\alpha}{j} x_{k-j+1} = A x_k + F x_k + \sum_{j=0}^{j=p} B_j u_{k-j} + \sum_{j=0}^{j=p} G_j u_{k-j}$$

where $n \times n$ dimensional matrix

$$F = \frac{d}{dx_k} f(x_k, u_k, u_{k-1}, \ldots, u_{k-j}, \ldots, u_{k-p})|_{0,0,0,\ldots,0}$$

and $n \times m$ dimensional matrices

$$G_j = \frac{d}{du_{k-j}} f(x_k, u_k, u_{k-1}, \ldots, u_{k-j}, \ldots, u_{k-p})|_{0,0,0,\ldots,0}$$

Moreover, for simplicity of notation let us denote

$$A + F = C, D_j = B_j + G_j, j = 0, 1, 2, \ldots, p.$$

Thus we have

$$x_{k+1} + \sum_{j=1}^{j=k+1} (-1)^j \binom{\alpha}{j} x_{k-j+1} = C x_k + \sum_{j=0}^{j=p} D_j u_{k-j} \qquad (10.2)$$

10.2 Fractional Semilinear Systems with Multiple Delays in Control

Lemma 10.1 Kaczorek [15] *let us assume that* $u_{-1} = 0$. *Hence, the solution of linear difference* Eq. (10.2) *with initial condition* $x_0 \in R^n$ *is given by*

$$x_{k+1} = \Phi_{k+1} x_0 + \sum_{i=0}^{i=k} H_{k-i} u_i \qquad (10.3)$$

where $n \times n$ *dimensional state transition matrices* Φ_k, $k = 0, 1, 2,...$ *are determined by the recurrent formula*

$$\Phi_{k+1} = (C + I_n \alpha)\Phi_k + \sum_{i=2}^{i=k+1} (-1)^{i+1} \binom{\alpha}{i} \Phi_{k-i+1} \qquad (10.4)$$

with $\Phi_0 = I$, where I is $n \times n$ dimensional identity matrix and by assumption matrices $\Phi_k = 0$ for $k < 0$, and $n \times m$ dimensional matrices H_k are defined as follows

$$H_0 = D_0$$

$$H_k = \sum_{i=0}^{i=k} \Phi_{k-i} D_i \quad for\ k = 1, 2, 3, \ldots, p - 1$$

$$H_k = \sum_{i=0}^{i=p} \Phi_{k-i} D_i \quad for\ k = p, p+1, p+2, \ldots$$

Moreover, it should be pointed out, that the matrices Φ_k, $k = 0, 1, 2,...$ defined above are extensions for fractional linear discrete-time control systems, the well known state transition matrices (see e.g. [4]) for standard linear discrete-time control systems.

10.3 Controllability Conditions

First of all, in order to define global and local controllability concepts for semilinear and linear finite-dimensional discrete-time control systems let us introduce the notion of reachable set or in other words attainable set in q steps [15, 33].

Definition 10.1 For fractional semilinear system (10.1) or linear system (10.2) reachable set in q steps from initial condition $x_0 = 0$ is defined as follows:

$K_q = \{x(q) \in R^n : x(q)\}$ is a solution of semilinear system (10.1) or linear system (10.2) in step q for sequence of admissible controls

$$u_0, u_1, \ldots u_k, \ldots u_{q-1}\} \qquad (10.5)$$

Definition 10.2. The fractional semilinear discrete-time control system (10.1) *is locally controllable in q-steps* if there exists a neighborhood of zero, $N \subset R^n$, such that

$$K_q = N \tag{10.6}$$

Definition 10.3 The fractional linear discrete-time linear control system (10.2) is globally controllable in q-steps if

$$K = R^n \tag{10.7}$$

For linear control system (10.2) let us introduce the $n \times 2qm$ dimensional controllability matrix

$$S_q = \begin{bmatrix} H_0 & | & H_1 & | & \cdots & | & H_k & | & \cdots & | & H_{q-1} \end{bmatrix} \tag{10.8}$$

In order to prove sufficient condition for local controllability of semilinear discrete-time fractional control systems (10.2), we shall use certain result taken directly from nonlinear functional analysis. This result concerns so called nonlinear covering operators.

Lemma 10.2 Robinson [68] *let $W: Z \to Y$ be a nonlinear operator from a Banach space Z into a Banach space Y and $W(0) = 0$. Moreover, it is assumed, that operator W has the Frechet derivative $dW(0): Z \to Y$, whose image coincides with the whole space Y. Then the image of the operator F will contain a neighborhood of the point $W(0) \in Y$.*

Now, we are in the position to formulate and prove the main result on the local unconstrained controllability in the interval [0, q] for the nonlinear discrete system (10.1). This result is known for semilinear or nonlinear continuous-time control system and is given in [7], as a sufficient condition for local controllability.

Theorem 10.1 *Semilinear discrete-time control system (10.1) is locally controllable in q steps if the associated linear discrete-time control system (10.2) is globally controllable in q-steps.*

Proof Proof of the Theorem 10.1 is based on Lemmas 10.1 and 10.2. Let our nonlinear operator W transforms the space of admissible control sequence $\{u(i) : 0 \le i \le q\}$ into the space of solutions at the step q for the semilinear discrete-time fractional control system (10.1).

More precisely, the nonlinear operator $W : R^m \times R^m \times \cdots \times R^m \to R^n$ associated with semilinear control system (10.1) is defined as follows [7]:

$$W\{u(0), u(1), u(2), \ldots, u(i), \ldots, u(q-1)\} = x_{sem}(q)$$

where $x_{sem}(q)$ is the solution at the step q of the semilinear discrete-time fractional control system (10.1) corresponding to an admissible controls sequence $u_q = \{u(i) : 0 \le i < q\}$.

10.3 Controllability Conditions

Therefore, for zero initial condition Frechet derivative at point zero of the nonlinear operator W denoted as $dW(0)$ is a linear bounded operator defined by the following formula

$$dW(0)\{u(0), u(1), u(2), \ldots, u(i), \ldots, u(q-1)\} = x_{lin}(q) \quad (10.9)$$

where $x_{lin}(q)$ is the solution at the step q of the linear system (10.2) corresponding to an admissible controls sequence $u_q = \{u(i) : 0 \leq i < q\}$ for zero initial condition.

Since from the assumption nonlinear function $f(0,0) = 0$, then for zero initial condition nonlinear operator W transforms zero in the space of admissible controls into zero in the state space, i.e. $W(0) = 0$.

Moreover, let us observe, that if the associated linear discrete-time fractional control system (10.2) is globally controllable in the interval [0, q], then by Definition 10.1 the image of the Frechet derivative $dW(0)$ covers whole state space R^n.

Therefore, by the result stated at the beginning of the proof, the nonlinear operator W covers some neighborhood of zero in the state space R^n. Hence, by Definition 10.2 semilinear discrete-time fractional control system (10.1) is locally controllable in the interval [0, q]. Hence our Theorem 10.1 follows.

Now, for the convenience, let us recall some well known (see e.g. [1–6]) facts from the controllability theory of linear finite-dimensional discrete-time fractional control systems.

Theorem 10.2 Kaczorek [15] *the fractional discrete-time linear system* (10.2) *is globally controllable in q steps if and only if*

$$\text{rank } S_q = n \quad (10.10)$$

Taking into account the form of controllability matrix, from Theorem 10.2 immediately follows simple Corollary.

Corollary 10.1 Kaczorek [15] *the fractional linear control system* (10.2) *is controllable in q steps if and only $n \times n$ dimensional constant matrix $C_q C_q^T$ is invertible, i.e., there exists inverse matrix $\left(C_q C_q^T\right)^{-1}$.*

Corollary 10.2 *The fractional semilinear control system* (10.1) *is locally controllable in q steps if equality* (10.9) *holds or equivalently if $n \times n$ dimensional constant matrix $C_q C_q^T$ is invertible, i.e., there exists inverse matrix $\left(C_q C_q^T\right)^{-1}$.*

10.4 Controllability of Discrete-Time Semilinear Systems with Delay in Control

Let us consider the special case of fractional discrete linear system (10.1), i.e. system with one delay in the control, described by the semilinear difference state-space equation

$$\Delta^\alpha x_{k+1} = A x_k + B_0 u_k + B_1 u_{k-1} + f(x_k, u_k, u_{k-1}) \quad (10.11)$$

where

$x_k \in R^n$, $u_k \in R^m$ are the state and input,
A, B_0, B_1 are $n \times n$ and $n \times m$ constant matrices,
$f: R^n \times R^m \times R^m \to R^n$ is nonlinear function differentiable near zero in the space
$R^n \times R^m \times R^m$ and such that $f(0, 0, 0) = 0$.

Let us observe, that semilinear discrete-time control system is described by the difference state equation, which contains both pure linear and pure nonlinear parts in the right hand side of the state equation.

Similarly as before, using definition of fractional difference we may write semilinear difference Eq. (10.11) in the equivalent form

$$x_{k+1} + \sum_{j=1}^{j=k+1} (-1)^j \binom{\alpha}{j} x_{k-j+1} = A x_k + B_0 u_k + B_1 u_{k-1} + f(x_k, u_k, u_{k-1})$$

Next, using standard linearization method [7] it is possible to find the associated linear difference state equation

$$x_{k+1} + \sum_{j=1}^{j=k+1} (-1)^j \binom{\alpha}{j} x_{k-j+1}$$
$$= A x_k + B_0 u_k + B_1 u_{k-1} + F x_k + G_0 u_k + G_1 u_{k-1}$$

where $n \times n$ dimensional matrix

$$F = \frac{d}{dx_k} f(x_k, u_k, u_{k-1}) \big|_{x_k=0, u_k=0, u_{k-1}=0}$$

and $n \times m$ dimensional matrices

$$G_0 = \frac{d}{du_k} f(x_k, u_k, u_{k-1}) \big|_{x_k=0, u_k=0, u_{k-1}=0}.$$

$$G_1 = \frac{d}{du_{k-1}} f(x_k, u_k, u_{k-1}) \big|_{x_k=0, u_k=0, u_{k-1}=0}.$$

Moreover, for simplicity of notation let us denote

$$A + F = C, \quad D_0 = B_0 + G_0, \quad D_1 = B_1 + G_1.$$

10.4 Controllability of Discrete-Time Semilinear Systems with Delay in Control

Thus we have

$$x_{k+1} + \sum_{j=1}^{j=k+1} (-1)^j \binom{\alpha}{j} x_{k-j+1} = Cx_k + D_0 u_k + D_1 u_{k-1} \qquad (10.12)$$

Lemma 10.3 Balachandran and Dauer [3] and Breuer and Petruccione [6] *let us assume that $u_1 = 0$. Hence, the solution of linear difference Eq. (10.2) with initial condition $x_0 \in R^n$ is given by*

$$x_{k+1} = \Phi_{k+1} x_0 + \sum_{i=0}^{i=k} H_{k-i} u_i \qquad (10.13)$$

where $n \times n$ dimensional state transition matrices Φ_k, $k = 0, 1, 2,\ldots$ are determined by the recurrent formula

$$\Phi_{k+1} = (C + I_n \alpha)\Phi_k + \sum_{i=2}^{i=k+1} (-1)^{i+1} \binom{\alpha}{i} \Phi_{k-i+1} \qquad (10.14)$$

with $\Phi_0 = I$, where I is $n \times n$ dimensional identity matrix and by assumption matrices $\Phi_k = 0$ for $k < 0$, and $n \times m$ dimensional matrices H_k are defined as follows

$$H_0 = D_0$$
$$H_k = \Phi_k D_0 + \Phi_{k-1} D_1 \quad for\ k = 1, 2, 3, \ldots$$

Moreover, it should be pointed out, that the matrices Φ_k, $k = 0, 1, 2,\ldots$ defined above are extensions for fractional linear discrete-time control systems, the well known state transition matrices (see e.g. [4]) for standard linear discrete-time control systems.

10.5 Controllability Conditions

First of all, let us observe that global and local controllability concepts for semilinear and linear finite-dimensional discrete-time control systems with delay in control introduce in Sect. 10.2 are of course valid for system (10.11) with one delay in control.

For linear control system (10.12) let us introduce the $n \times qm$ dimensional controllability matrix

$$S_q = \begin{bmatrix} H_0 & | & H_1 & | & \cdots & | & H_k & | & \cdots & | & H_{q-1} \end{bmatrix} \qquad (10.15)$$

Now, we are in the position to formulate and prove the main result on the local unconstrained controllability in the interval [0, q] for the nonlinear discrete system (10.11). This result is known for semilinear or nonlinear continuous-time control system and is given in [7], as a sufficient condition for local controllability.

Theorem 10.3 *Semilinear discrete-time control system (10.11) is locally controllable in q steps if the associated linear discrete-time control system (10.12) is globally controllable in q-steps.*

Proof Proof of the Theorem 10.3 is based on Lemmas 10.1 and 10.2. Let our nonlinear operator W transforms the space of admissible control sequence $\{u(i): 0 \leq i \leq q\}$ into the space of solutions at the step q for the semilinear discrete-time fractional control system (10.11).

More precisely, the nonlinear operator

$$W : R^m \times R^m \times \ldots R^m \to R^n$$

associated with semilinear control system (10.11) is defined as follows [7]:

$$W\{u(0), u(1), u(2), \ldots, u(i), \ldots, u(q-1)\} = x_{sem}(q)$$

where $x_{sem}(q)$ is the solution at the step q of the semilinear discrete-time fractional control system (10.11) corresponding to an admissible controls sequence $u_q = \{u(i): 0 \leq i < q\}$.

Therefore, for zero initial condition Frechet derivative at point zero of the nonlinear operator W denoted as $dW(0)$ is a linear bounded operator defined by the following formula

$$dW(0)\{u(0), u(1), u(2), \ldots, u(i), \ldots, u(q-1)\} = x_{lin}(q)$$

where $x_{lin}(q)$ is the solution at the step q of the linear system (10.12) corresponding to an admissible controls sequence $u_q = \{u(i): 0 \leq i < q\}$ for zero initial condition.

Since from the assumption nonlinear function $f(0, 0) = 0$, then for zero initial condition nonlinear operator W transforms zero in the space of admissible controls into zero in the state space, i.e., $W(0) = 0$.

Moreover, let us observe, that if the associated linear discrete-time fractional control system (10.12) is globally controllable in the interval [0, q], then by Definition 10.1 the image of the Frechet derivative $dW(0)$ covers whole state space R^n.

Therefore, by the result stated at the beginning of the proof, the nonlinear operator W covers some neighborhood of zero in the state space R^n. Hence, by Definition 10.2 semilinear discrete-time fractional control system (10.11) is locally controllable in the interval [0, q]. Hence our Theorem 10.1 follows.

Now, for the convenience, let us recall some well known (see e.g. [1–6]) facts from the controllability theory of linear finite-dimensional discrete-time fractional control systems.

10.5 Controllability Conditions

Theorem 10.4 Breuer and Petruccione [6] *the fractional discrete-time linear system* (10.12) *is globally controllable in q steps if and only if*

$$\text{rank } S_q = n \tag{10.16}$$

Taking into account the form of controllability matrix, from Theorem 10.2 immediately follows simple Corollary.

Corollary 10.3 Breuer and Petruccione [6] *the fractional linear control system* (10.12) *is controllable in q steps if and only $n \times n$ dimensional constant matrix $C_q C_q^T$ is invertible, i.e., there exists inverse matrix* $\left(C_q C_q^T\right)^{-1}$.

Corollary 10.4 *The fractional semilinear control system* (10.11) *is locally controllable in q steps if equality* (10.19) *holds or equivalently if $n \times n$ dimensional constant matrix $C_q C_q^T$ is invertible, i.e., there exists inverse matrix* $\left(C_q C_q^T\right)^{-1}$.

Example 10.1 Let us consider as the illustrative example, semilinear fractional discrete-time control system with delay in control and constant coefficients of the form (10.11). It is assumed, that system parameters are $0 \leq \alpha \leq 1$, $n = 2$, $m = 1$, and $q = 2$. Moreover, matrices, vectors and functions in the difference state equation are given by the following equalities:

$$A = \begin{bmatrix} 1 & 0 \\ 0 & 1 \end{bmatrix} \quad B_0 = \begin{bmatrix} 0 \\ 1 \end{bmatrix} \quad B_1 = \begin{bmatrix} 0 \\ 1 \end{bmatrix}$$

$$f(x_k, u_k, u_{k-1}) = \begin{bmatrix} e^{u_{k-1}} - 1 \\ 2 \sin x_{k1} \end{bmatrix} \tag{10.17}$$

Hence we have

$$f(0,0,0) = \begin{bmatrix} 0 \\ 0 \end{bmatrix}$$

$$F = \frac{d}{dx_k} f(x_k, u_k, u_{k-1})_{|x_k=0, u_k=0, u_{k-1}=0} = \begin{bmatrix} 0 & 0 \\ 2 & 0 \end{bmatrix}$$

$$G_0 = \frac{d}{du_k} f(x_k, u_k, u_{k-1})_{|x_k=0, u_k=0, u_{k-1}=0} = \begin{bmatrix} 0 \\ 0 \end{bmatrix}$$

$$G_1 = \frac{d}{du_{k-1}} f(x_k, u_k, u_{k-1})_{|x_k=0, u_k=0, u_{k-1}=0} = \begin{bmatrix} 1 \\ 0 \end{bmatrix}$$

Therefore, matrices and vectors in the linear associated control system are given by equalities

$$C = A + F = \begin{bmatrix} 1 & 0 \\ 2 & 1 \end{bmatrix}$$

$$D_0 = B_0 + G_0 = \begin{bmatrix} 0 \\ 1 \end{bmatrix}$$

$$D_1 = B_1 + G_1 = \begin{bmatrix} 1 \\ 1 \end{bmatrix}$$

Using formula (10.4) for $k = 0$ we obtain transition state matrix for linear fractional system

$$\Phi_1 = (C + I\alpha)\Phi_0 = \begin{bmatrix} 1+\alpha & 0 \\ 2 & 1+\alpha \end{bmatrix}$$

Controllability matrix (10.8) S_2 for $q = 2$ has the form

$$S_2 = [H_0 \mid H_1] = [D_0 \mid \Phi_1 D_0 + D_1] = [B_0 + G_0 \mid \Phi_1(B_0 + G_0) + (B_1 + G_1)]$$

Hence, taking into account the matrices calculated above we have:

$$S_2 = \begin{bmatrix} 0 & 1 \\ 1 & 2+\alpha \end{bmatrix}$$

Therefore, since *rank* $S_2 = 2 = n$, for each coefficient α, then taking into account Theorem 10.2 or Theorem 10.4 the fractional associated linear discrete-time system with constant coefficients is globally controllable in two steps, hence by Theorem 10.1 or Theorem 10.3 the semilinear fractional discrete-time system (10.17) is locally controllable in two steps for each coefficient α.

For comparison let us consider linear fractional discrete system (10.2) with the matrices A, B_0, B_1 given by equalities (10.17). In this case using formula (10.4) for $k = 0$ we have

$$\Phi_1 = (A + I\alpha)\Phi_0 = \begin{bmatrix} 1+\alpha & 0 \\ 0 & 1+\alpha \end{bmatrix}$$

Controllability matrix (10.8) S_2 for $q = 2$ has the form

$$S_2 = [B_0 \mid \Phi_1 B_0 + B_1] = \begin{bmatrix} 0 & 0 \\ 1 & 2+\alpha \end{bmatrix}$$

Therefore, since *rank* $R_2 = 1 < n$ then, taking into account Corollary 10.1 or Corollary 10.3 the fractional linear discrete-time system (10.17) with single delay and constant coefficients is not globally controllable in two steps and consequently in any number of steps for each coefficient α.

10.5 Controllability Conditions

In the present Chapter unconstrained local controllability problem of finite-dimensional fractional-discrete time semilinear systems with delay in control and constant coefficient is addressed. Using linearization method and solution formula for linear fractional difference equation with delay in control sufficient condition for unconstrained local controllability in q steps of the discrete time fractional control system has been established as rank condition of suitably defined controllability matrix.

In the proof of the main result certain theorem taken directly from nonlinear functional analysis is used. Moreover, simple illustrative numerical example is also presented.

There are many possible extensions of the results given in the paper. First of all it is possible to consider semilinear infinite-dimensional fractional control systems. In this case it it necessary to introduce two fundamental concepts of controllability, namely: approximate (weak) controllability and exact (strong) controllability.

Moreover, it should be mentioned, that controllability considerations presented in the chapter can be extended for fractional discrete-time linear systems with multiple delays both in the controls and in the state variables. Finally, all the results may be extended to cover the cases of fractional linear or semilinear discrete time control systems with variable coefficients.

Chapter 11
Stochastic Controllability and Minimum Energy Control of Systems with Multiple Variable Delays in Control

11.1 Introduction

In recent years various controllability problems for different types of linear dynamical systems have been considered in many publications and monographs. The extensive list of these publications can be found for example in the monograph [31] or in the survey papers [32–34].

However, it should be stressed, that the most literature in this direction has been mainly concerned with deterministic controllability problems for finite-dimensional linear dynamical systems with unconstrained controls and without delays.

For stochastic control systems both linear and nonlinear the situation is less satisfactory. In recent years the extensions of the deterministic controllability concepts to stochastic control systems have been recently discussed only in a rather few number of publications. Lyapunov techniques were used to formulate and prove sufficient conditions for stochastic controllability of nonlinear finite dimensional stochastic systems with point delays in the state variable. Moreover, it should be pointed out, that the functional analysis approach to stochastic controllability problems is also extensively discussed both for linear and nonlinear stochastic control systems.

In the present Chapter we shall study stochastic controllability problems for linear dynamical systems, which are natural generalizations of controllability concepts well known in the theory of infinite dimensional control systems [8, 31–33, 38, 39, 45, 47, 55, 56]. More precisely, we shall consider stochastic relative exact and approximate controllability problems for finite-dimensional linear non-stationary dynamical systems with multiple time-variable point delays in the control described by stochastic ordinary differential state equations. Hence, using techniques similar to those presented in the papers [61, 62] we shall formulate and prove necessary and sufficient conditions for stochastic relative exact controllability in a prescribed time interval for linear nonstationary stochastic dynamical systems with multiple time-variable point delays in the control.

Roughly speaking, it will be proved that under suitable assumptions relative controllability of a deterministic linear associated dynamical system is equivalent to stochastic relative exact controllability and stochastic relative approximate controllability of the original linear stochastic dynamical system. This is a generalization to control delayed case some previous results concerning stochastic controllability of linear dynamical systems without delays in control [19, 61, 62].

The chapter is organized as follows: Sect. 11.2 contains mathematical model of linear, nonstationary stochastic dynamical system with multiple time-variable point delays in the control. Moreover, in this section basic notations and definitions of stochastic relative exact and stochastic approximate controllability are presented and some preliminary results are also included.

In Sect. 11.3 using results and methods taken directly from deterministic controllability problems, necessary and sufficient conditions for exact and approximate stochastic relative controllability are formulated and proved.

In Sect. 11.4, as a special case relative stochastic controllability in a given time interval of dynamical systems with multiple constant point delays is also considered.

Sections 11.5 and 11.6 are devoted to a study of controllability problem for systems both with single time-variable and constant delay in control, respectively.

In Sect. 11.7 minimum energy control problem for general nonstationary and stationary systems with multiple delays is formulated and analytically solved using methods taken directly from functional analysis.

In Sect. 11.8 minimum energy control problem for nonstationary and stationary systems both with single time-variable and constant delay in control is studied. Finally, concluding remarks and states some open controllability problems for more general stochastic dynamical systems are given at the end of chapter.

11.2 System Description

Throughout this Chapter, unless otherwise specified, we use the following standard notations. Let (Ω, F, P) be a complete probability space with probability measure P on Ω and a filtration $\{F_t | t \in [t_0, T]\}$ generated by n-dimensional Wiener process $\{w(s): t_0 \leq s \leq t\}$ defined on the probability space (Ω, F, P).

Let $L_2(\Omega, F_T, R^n)$ denotes the Hilbert space of all F_T-measurable square integrable random variables with values in R^n. Moreover, let $L_2^F([t_0, T], R^n)$ denotes the Hilbert space of all square integrable and F_t-measurable processes with values in R^n.

In the theory of linear, finite-dimensional, time-invariant stochastic dynamical control systems we use mathematical model given by the following stochastic ordinary differential state equation with multiple time-variable point delays in the control.

11.2 System Description

$$dx(t) = (A(t)x(t) + \sum_{i=0}^{i=M} B_i(t)u(v_i(t)))dt + \sigma(t)dw(t) \quad \text{for} \quad t \in [t_0, T], \quad (11.1)$$

with initial conditions:

$$x(t_0) = x_0 \in L_2(\Omega, F_T, R^n)$$

$$\text{and } u(t) = 0 \quad \text{for } t \in [v_M(t_0), t_0) \quad (11.2)$$

where the state $x(t) \in R^n = X$, the control $u(t) \in R^m = U$,
$A(t)$ is $n \times n$ dimensional matrix with elements $a_{kj} \in L_2([t_0, T])$ for $k, j = 1, 2, \ldots, n$,
$B_i(t)$, $i = 0, 1, 2,\ldots M$, are $n \times m$ dimensional matrices with elements $b_{ikj} \in L_2([t_0, T])$ for $j = 1, 2,\ldots, m$, $k = 1, 2,\ldots n$, $\sigma(t)$ is $n \times n$ dimensional matrix with continuous elements, and $v_i(t)$, $i = 1, 2,\ldots, M$ are time-variable point delays.

It is generally assumed that the real valued functions $v_i(t)$, $t \in [t_0, T]$ are absolutely continuous and strictly increasing and satisfy the following inequalities:

$$v_M(t) < v_{M-1}(t) < \cdots < v_i(t) < \cdots < v_1(t) < v_0(t) = t \quad \text{for} \quad t \in [t_0, T]$$

It should be mentioned that functions $v_i(t)$, $i = 1, 2,\ldots, M$ can also be expressed in the following frequently used form:

$$v_i(t) = t - h_i(t) \quad \text{for } i = 1, 2, \ldots, M, \text{ and } t \in [t_0, T]$$

where $h_i(t) > 0$, for $i = 1, 2,\ldots, M$, and $t \in [t_0, T]$ are time-variable point delays.
Hence, we have the following inequalities

$$0 = h_0(t) < h_1(t) < \cdots < h_i(t) < \cdots < h_{M-1}(t) < h_M(t) \quad \text{for } t \in [t_0, T]$$

In order to simplify the notation, let us introduce for every function $v_i(t)$ so called leading function $r_i(t)$, which is simply the inverse function for $v_i(t)$. Hence, the leading function $r_i(t)$: $[v_i(t_0), v_i(T)] \rightarrow [t_0, T]$ satisfy the following equality.

$$r_i(v_i(t)) = t \quad \text{for } i = 1, 2, \ldots, M, \text{ and } t \in [t_0, T]$$

Therefore, we have

$$t = r_0(t) < r_1(t) < \cdots < r_i(t) < \cdots < r_{M-1}(t) < r_M(t) \quad \text{for } t \in [t_0, T]$$

In the sequel for simplicity of considerations we generally assume that the set of admissible controls $U_{ad} = L_2^F([t_0, T], R^m)$.

It is well known (see e.g. [31, 32, 61], or [62] that for a given initial conditions (11.2) and any admissible control $u \in U_{ad}$, for $t \in [t_0, T]$ there exist unique solution $x(t; x_0, u) \in L_2(\Omega, F_t, R^n)$ of the linear stochastic differential state Eq. (11.1) which can be represented in the following integral form

$$x(t; x_0, u) = F(t, t_0)x_0 + \int_{t_0}^{t} F(t, s) \left(\sum_{i=0}^{i=M} B_i(s)u(v_i(s)) \right) ds$$

$$+ \int_{t_0}^{t} F(t, s)\sigma(s)dw(s)$$

where $F(t, s)$ is n × n dimensional well-known so called state transition matrix, generated by the matrix $A(t)$, (see e.g., [31] for properties of $F(t, s)$ and more details).

Thus, without loss of generality, taking into account zero initial control for $t \in [v_M(t_0), t_0)$, and changing the order of integration, the solution $x(t; x_0, u)$ for $r_k(t_0) < t \leq r_{k+1}(t_0)$, $k = 0, 1, 2,\ldots, M - 1$, has the following form, which is more convenient for further considerations [31].

$$x(t; x_0, u) = F(t, t_0)x_0$$

$$+ \sum_{i=0}^{i=k} \int_{t_0}^{v_i(t)} F(t, r_i(s))B_i(r_i(s))r'_i(s)u(s)ds$$

$$+ \int_{t_0}^{t} F(t, s)\sigma(s)dw(s)$$

Moreover, for $t > r_M(t_0)$ we have

$$x(t; x_0, u) = F(t, t_0)x_0$$

$$+ \sum_{i=0}^{i=M} \int_{t_0}^{v_i(t)} F(t, r_i(s))B_i(r_i(s))r'_i(s)u(s)ds$$

$$+ \int_{t_0}^{t} F(t, s)\sigma(s)dw(s)$$

11.2 System Description

or equivalently for $r_k(t_0) < t \leq r_{k+1}(t_0)$, $k = 0, 1, 2, \ldots, M-1$,

$$x(t; x_0, u) = \exp(At)x_0$$
$$+ \sum_{i=0}^{i=k-1} \int_{v_{i+1}(t)}^{v_i(t)} \left(\sum_{j=0}^{j=i} F(t, r_j(s)) B_j(r_j(s)) r'_j(s) \right) u(s) ds$$
$$+ \int_{t_0}^{v_k(t)} \left(\sum_{j=0}^{j=k} F(t, r_j(s)) B_j(r_j(s)) r'_j(s) \right) u(s) ds$$
$$+ \int_{t_0}^{t} F(t, s) \sigma(s) dw(s)$$

and similarly for $t > r_M(t_0)$,

$$x(t; x_0, u) = F(t, t_0)x_0$$
$$+ \sum_{i=0}^{i=k-1} \int_{v_{i+1}(t)}^{v_i(t)} \left(\sum_{j=0}^{j=i} F(t, r_j(s)) B_j(r_j(s)) r'_j(s) \right) u(s) ds$$
$$+ \int_{t_0}^{v_M(t)} \left(\sum_{j=0}^{j=M} F(t, r_j(s)) B_j(r_j(s)) r'_j(s) \right) u(s) ds$$
$$+ \int_{t_0}^{t} F(t, s) \sigma(s) dw(s)$$

Now, for a given final time T, using the form of the integral solution $x(t; x_0, u)$ let us introduce operators and sets, which will be used in next parts of the paper [31].

First of all for $r_k(t_0) < T \leq r_{k+1}(t_0)$, $k = 0, 1, 2, \ldots, M-1$, we define linear and bounded control operator $L_T \in L_2^F([0, T], R^m) \to L_2(\Omega, F_T, R^n)$ as follows:

$$L_T u = \sum_{i=0}^{i=k-1} \int_{v}^{v} \left(\sum_{j=0}^{j=i} F(T, r_j(s)) B_j(r_j(s)) r'_j(s) \right) u(s) ds$$
$$+ \int_{t_0}^{v_k(T)} \left(\sum_{j=0}^{j=k} F(T, r_j(s)) B_j(r_j(s)) r'_j(s) \right) u(s) ds$$

Moreover, for $T > r_M(t_0)$ we have:

$$L_T u = \sum_{i=0}^{i=M-1} \int_{v_{i+1}(T)}^{v_i(T)} \left(\sum_{j=0}^{j=i} F(T, r_j(s)) B_j(r_j(s)) r'_j(s) \right) u(s) ds$$

$$+ \int_{t_0}^{v_M(T)} \left(\sum_{j=0}^{j=M} F(T, r_j(s)) B_j(r_j(s)) r'_j(s) \right) u(s) ds$$

and its adjoint linear and bounded operator

$$L_T^* \in L_2(\Omega, F_T, R^n) \to L_2^F([0, T], R^m)$$

$$L_T^* z = B_0^*(t) F^*(T, t)) E\{z | F_t\} \quad \text{for } t \in [t_0, v_M(T)] L_T^* z$$

$$= \left(\sum_{j=1}^{j=i} B_j^*(t) F^*(T, r_j(t)) r'_j(t) \right) E\{z | F_t\}$$

for $t \in (v_i(T), v_{i+1}(T)] i = 0, 1, \ldots, M - 1$

From the above notation it follows that the set of all states reachable from initial state $x(t_0) = x_0 \in L_2(\Omega, F_T, R^n)$ in time $T > t_0$, using admissible controls has the form

$$R_T(U_{ad}) = \{x(T; x_0, u) \in L_2(\Omega, F_T, R^n) : u \in U_{ad}\}$$

$$= F(T, t_0) x_0 + \text{Im } L_T + \int_{t_0}^{T} F(T, s) \sigma(s) dw(s)$$

Moreover, we introduce the concept of the linear controllability operator [8, 9, 14, 19] $C_T \in L(L_2(\Omega, F_T, R^n), L_2(\Omega, F_T, R^n))$ which is strongly associated with the control operator L_T and is defined by the following equality

$$C_T = L_T L_T^* = \sum_{i=0}^{i=k-1} \int_{v_{i+1}(T)}^{v_i(T)} P(s) P^*(s) E\{\cdot | F_t\} dt$$

$$+ \int_{t_0}^{v_k(T)} P(s) P^*(s) E\{\cdot | F_t\} dt$$

11.2 System Description

where

$$P(s) = \left(\sum_{j=0}^{j=i} F(T, r_j(s)) B_j(r_j(s)) r'_j(s) \right)$$

$$P^*(s) = \left(\sum_{j=0}^{j=i} B^*_j(r_j(s)) F^*(T, r_j(s)) r'_j(s) \right)$$

for $r_i(t_0) < T \leq r_{i+1}(t_0)$, $i = 0, 1, \ldots, M - 1$, and for $T > r_M(t_0)$ by the equality

$$C_T = L_T L_T^* = \sum_{i=0}^{i=M-1} \int_{v_{i+1}(T)}^{v_i(T)} P(s) P^*(s) E\{\cdot | F_t\} dt + \int_{t_0}^{v_M(T)} P(s) P^*(s) E\{\cdot | F_t\} dt$$

Finally, let us recall n × n-dimensional deterministic controllability matrix [31] for $r_i(t_0) < T \leq r_{i+1}(t_0)$, $i = 0, 1, \ldots, M - 1$

$$G_T = L_T L_T^* = \sum_{i=0}^{i=k-1} \int_{v_{i+1}(T)}^{v_i(T)} P(s) P^*(s) dt + \int_{t_0}^{v_k(T)} P(s) P^*(s) dt$$

and for $T > r_M(t_0)$

$$G_T = L_T L_T^* = \sum_{i=0}^{i=M-1} \int_{v_{i+1}(T)}^{v_i(T)} P(s) P^*(s) dt + \int_{t_0}^{v_M(T)} P(s) P^*(s) dt$$

In the proofs of the main results we shall use also controllability operators $C_T(s)$ and controllability matrices $G_T(s)$ depending on time and defining as follows, for $r_k(t_0) < T < r_k(t_0)$, $k = 0, 1, \ldots, M - 1$,

$$C_T(s) = L_T L_T^* = \sum_{i=0}^{i=k-1} \int_{v_{i+1}(T)}^{v_i(T)} P(s) P^*(s) E\{\cdot | F_t\} dt$$

$$+ \int_{t_0}^{v_k(T)} P(s) P^*(s) E\{\cdot | F_t\} dt$$

and for $T > h_M$

$$C_T(s) = L_T L_T^* = \sum_{i=0}^{i=k-1} \int_{v_{i+1}(T)}^{v_i(T)} P(s)P^*(s)E\{\cdot|F_t\}dt$$
$$+ \int_{t_0}^{v_M(T)} P(s)P^*(s)E\{\cdot|F_t\}dt$$

Similarly for $r_k(t_0) < T < r_{k+1}(t_0)$, $k = 0, 1, \ldots, M - 1$, we have

$$G_T(s) = L_T L_T^* = \sum_{i=0}^{i=k-1} \int_{v_{i+1}(T)}^{v_i(T)} P(s)P^*(s)dt$$
$$+ \int_{s}^{v_k(T)} P(s)P^*(s)dt$$

and for $T > r_M(t_0)$

$$G_T(s) = L_T L_T^* = \sum_{i=0}^{i=k-1} \int_{v_{i+1}(T)}^{v_i(T)} P(s)P^*(s)dt$$
$$+ \int_{s}^{v_M(T)} P(s)P^*(s)dt$$

It is well known, that in the theory of dynamical systems with delays in control or in the state variables, it is necessary to distinguish between two fundamental concepts of controllability, namely: relative controllability and absolute controllability, (see e.g. [31, 32], or [34] for more details). In this paper we shall concentrate on the weaker concept relative controllability on a given time interval $[t_0, T]$.

On the other hand, since for the stochastic dynamical system (11.1) the state space $L_2(\Omega, F_t, R^n)$ is in fact infinite-dimensional space, we distinguish exact or strong controllability and approximate or weak controllability.

Using the notations given above for the stochastic dynamical system (11.1) we define the following stochastic relative exact and approximate controllability concepts.

Definition 11.1 The stochastic dynamical system (11.1) is said to be stochastically relatively exactly controllable on $[t_0, T]$ if $R_T(U_{ad}) = L_2(\Omega, F_T, R^n)$ that is, if all the

11.2 System Description

points in $L_2(\Omega, F_T, R^n)$ can be exactly reached from arbitrary initial condition $x_0 \in L_2(\Omega, F_T, R^n)$ at time T.

Definition 11.2 The stochastic dynamical system (11.1) is said to be stochastically relatively approximately controllable on $[t_0, T]$ if $\overline{R_T(U_{ad})} = L_2(\Omega, F_T, R^n)$ that is, if all the points in $L_2(\Omega, F_T, R^n)$ can be approximately reached from arbitrary initial condition $x_0 \in L_2(\Omega, F_T, R^n)$ at time T.

Remark 11.1 From the Definitions 11.1 and 11.2 directly follows, that stochastic relative exact controllability is generally a stronger concept than stochastic relative approximate controllability. However, there are many cases when these two concepts coincide.

Remark 11.2 Since the stochastic dynamical system (11.1) is linear, then without loss of generality in the above two definitions it is enough to take zero initial condition $x_0 = 0 \in L_2(\Omega, F_T, R^n)$.

Remark 11.3 It should be pointed out, that in the case of delayed controls the above controllability concepts essentially depend on the length of the time interval $[t_0, T]$.

Remark 11.4 Let us observe, that for the final time $T \leq r_1(t_0)$ stochastic dynamical system (11.1) is in fact a system without delay.

Remark 11.5 Since the dynamical system (11.3) is stationary, therefore controllability matrix $G_T(s)$ has the same rank at least for all $s \in [t_0, v_k(T)]$, if $r_{k+1}(t_0) < T < r_k(t_0)$, $k = 0, 1, \ldots, M - 1$, or for $s \in [t_0, v_M(T)]$, if $T > r_M(t_0)$, [8, Chap. 4].

Remark 11.6 From the form of the controllability operator C_T immediately follows, that this operator is self adjoint.

In the sequel we study the relationship between the controllability concepts for the stochastic dynamical system (11.1) and controllability of the associated deterministic dynamical system of the form

$$y'(t) = A(t)y(t) + \sum_{i=0}^{i=M} B_i(t)w(v_i(t)) \quad t \in [t_0, T] \tag{11.3}$$

where the admissible controls $w \in L_2([t_0, T], R^m)$.

Therefore, let us recall the lemma concerning relative controllability of deterministic system (11.3).

Lemma 11.1 [31]. *The following conditions are equivalent:*

(i) *deterministic system (11.3) is relatively controllable on $[t_0, T]$,*
(ii) *controllability matrix G_T is nonsingular,*

Now, let us formulate auxiliary well known lemma taken directly from the theory of stochastic processes, which will be used in the sequel in the proofs of the main results.

Lemma 11.2 [14, 19, 20]. *For every*

$$z \in L_2(\Omega, F_T, R^n),$$

there exists a process $q \in L_2^F([0,T], R^{n \times n})$ *such that*

$$C_T z = G_T E z + \int_{t_0}^{T} G_T(s) q(s) dw(s)$$

Taking into account the above notations, definitions and lemmas in the next section we shall formulate and prove conditions for stochastic relative exact and stochastic relative approximate controllability for stochastic dynamical system (11.1).

11.3 Stochastic Relative Controllability

In this section, using lemmas given in Sect. 11.2 we shall formulate and prove the main result of the paper, which says that stochastic relative exact and in consequence also approximate controllability of stochastic system (11.1) is in fact equivalent to relative controllability of associated linear deterministic system (11.3).

Theorem 11.1 *The following conditions are equivalent:*

(i) *Deterministic system (11.3) is relatively controllable on* $[t_0, T]$,
(ii) *Stochastic system (11.1) is stochastically relatively exactly controllable on* $[t_0, T]$
(iii) *Stochastic system (11.1) is stochastically relatively approximately controllable on* $[t_0, T]$.

Proof (i) $\Rightarrow$ (ii) Let us assume that the deterministic system (11.3) is relatively controllable on $[t_0, T]$. Then, it is well known (see e.g. [8, 9], or [12]) that the symmetric relative deterministic controllability matrix $G_T(s)$ is invertible and strictly positive definite at least for all [31]

$$s \in [t_0, v_k(T)], \text{ if } h_k < T < h_{k+1}, \ k = 0, 1, 2, \ldots, M-1$$

or for at least for all [31]

$$s \in [t_0, v_M(T)] \text{if } T > v_M(T),$$

11.3 Stochastic Relative Controllability

Hence, for some $\gamma > 0$ we have

$$\langle G_T(s)x, x \rangle \geq \gamma \|x\|^2$$

for all $s \in [0, T - h_M]$ and for all $x \in R^n$. In order to prove stochastic relative exact controllability on $[0, T]$ for the stochastic system (11.1) we use the relationship between controllability operator C_T and controllability matrix G_T given in Lemma 11.2, to express $E\langle C_T z, z \rangle$ in terms of $\langle G_T Ez, Ez \rangle$. First of all we obtain

$$E\langle C_T z, z \rangle = E\left\langle G_T Ez + \int_{t_0}^{T} G_T(s)q(s)dw(s), Ez + \int_{t_0}^{T} q(s)dw(s) \right\rangle$$

$$= \langle G_T Ez, Ez \rangle$$

$$+ E \int_{t_0}^{T} \langle G_T(s)q(s), q(s) \rangle ds \geq \gamma \left(\|Ez\|^2 + E \int_{t_0}^{T} \|q(s)\|^2 ds \right) = \gamma E\|z\|^2$$

Hence, in the operator sense we have the following inequality $C_T \geq \gamma I$, which means that the operator C_T is strictly positive definite and thus, that the inverse linear operator C_T^{-1} is bounded. Therefore, stochastic relative exact controllability on $[0, T]$ of the stochastic dynamical system (11.1) directly follows from the results given in [30].

(ii) $\Rightarrow$ (iii) This implication is obvious (see e.g. [31, 61–63]).

(iii) $\Rightarrow$ (i) Assume that the stochastic dynamical system (11.1) is stochastically relatively approximately controllable on $[t_0, T]$, and hence its linear self adjoint controllability operator is positive definite, i.e. $C_T > 0$ [31].

Then, using the resolvent operator $R(\lambda, C_T)$ and following directly the functional analysis method given in [61–63] for stochastic dynamical systems without delays, we obtain that deterministic system (11.3) is approximately relatively controllable on $[t_0, T]$.

However, taking into account that the state space for deterministic dynamical system (11.3) is finite dimensional, i.e., exact and approximate controllability coincide [31], we conclude that deterministic dynamical system (11.3) is relatively controllable on $[t_0, T]$.

Remark 11.7 Let us observe, that for a special case when the final time $T \leq r_1(t_0)$, stochastic relative exact or approximate controllability problems in $[t_0, T]$ for stochastic dynamical system with delay in the control (11.1) are reduced to the standard stochastic exact or stochastic approximate controllability problems for the stochastic dynamical system without delays in the control [31].

Corollary 11.1 [61]. *Stochastic dynamical system without delays, i.e. $B_j = 0$, $j = 1, 2, \ldots, M$, is stochastically exactly controllable in time interval $[t_0, T]$ if and only if associated deterministic dynamical system without delay is controllable in time interval $[t_0, T]$.*

Remark 11.8 Finally, it should be pointed out, that using general method given in the monograph [30], for stochastically relatively approximately controllable dynamical systems (11.1) it is possible to formulate the admissible controls $u(t)$, defined for $t \in [t_0, T]$ and transferring given initial state x_0 to the desired final state x_T at time T with minimum energy.

11.4 Stationary Systems with Multiple Constant Delays

Now, as a special case we shall consider linear, finite-dimensional, time-invariant stochastic dynamical control systems given by the following stochastic ordinary differential state equation with multiple constant point delays in the control

$$dx(t) = (Ax(t) + \sum_{i=0}^{i=M} B_i u(t - h_i))dt + \sigma dw(t) \quad \text{for } t \in [0, T], \tag{11.4}$$

with initial conditions:

$$x(0) = x_0 \in L_2(\Omega, F_T, R^n) \text{ and } u(t) = 0 \quad \text{for } t \in [-h_M, 0) \tag{11.5}$$

where the state $x(t) \in R^n = X$ and the control $u(t) \in R^m = U$, A is $n \times n$ dimensional constant matrix, B_i, $i = 0, 1, 2,...M$, are $n \times m$ dimensional constant matrices, σ is $n \times n$ dimensional constant matrix, and

$$0 = h_0 < h_1 < \cdots < h_i < \cdots < h_{M-1} < h_M$$

are constant delays.

In this time invariant special case

$$v_i(t) = t - h_i, \quad i = 0, 1, 2, \ldots, M,$$

and the leading functions $r_i(t)$, $i = 0, 1, 2,..., M$ are as follows

$$r_i(t) = t + h_i \quad \text{for } i = 0, 1, 2, \ldots, M$$

In the sequel we assume that the set of admissible controls $U_{ad} = L_2^F([0, T], R^m)$.

Then for a given initial conditions (11.5) and any admissible control $u \in U_{ad}$, for $t \in [t_0, T]$ there exist unique solution

$$x(t; x_0, u) \in L_2(\Omega, F_t, R^n)$$

11.4 Stationary Systems with Multiple Constant Delays

of the linear stochastic differential state Eq. (11.4) which can be represented in the following integral form

$$x(t; x_0, u) = \exp(At)x_0 + \int_0^t \exp(A(t-s))\left(\sum_{i=0}^{i=M} B_i u(s - h_i)\right)ds$$

$$+ \int_0^t \exp(A(t-s))\sigma dw(s)$$

Thus, without loss of generality, taking into account zero initial control for $t \in [-h_M, 0]$, and changing the order of integration, the solution $x(t; x_0, u)$ for $h_k < t \leq h_{k+1}$, $k = 0, 1, 2, \ldots, M - 1$, $t \in [0, h]$ has the following form, which is more convenient for further considerations [31]

$$x(t; x_0, u) = \exp(At)x_0 + \sum_{i=0}^{i=k} \int_0^{t-h_i} \exp(A(t - s - h_i))B_i u(s)ds$$

$$+ \int_0^t \exp(A(t-s))\sigma dw(s)$$

Moreover, for $t > h_M$ we have

$$x(t; x_0, u) = \exp(At)x_0 + \sum_{i=0}^{i=M} \int_0^{t-h_i} \exp(A(t - s - h_i))B_i u(s)ds$$

$$+ \int_0^t \exp(A(t-s))\sigma dw(s)$$

or equivalently for $h_k < t < h_{k+1}$, $k = 0, 1, 2, \ldots, M - 1$,

$$x(t; x_0, u) = \exp(At)x_0$$

$$+ \sum_{i=0}^{i=k-1} \int_{t-h_{i+1}}^{t-h_i} \left(\sum_{j=0}^{j=i} \exp(A(t - s - h_j))B_j\right) u(s)ds$$

$$+ \int_0^{t-h_k} \left(\sum_{j=0}^{j=k} \exp(A(t - s - h_j))B_j\right) u(s)ds + \int_0^t \exp(A(t-s))\sigma dw(s)$$

and similarly for $t > h_M$,

$$x(t; x_0, u) = \exp(At)x_0$$
$$+ \sum_{i=0}^{i=k-1} \int_{t-h_{i+1}}^{t-h_i} \left(\sum_{j=0}^{j=i} \exp(A(t-s-h_j))B_j\right) u(s)ds$$
$$+ \int_{0}^{t-h_M} \left(\sum_{j=0}^{j=M} \exp(A(t-s-h_j))B_j\right) u(s)ds$$
$$+ \int_{0}^{t} \exp(A(t-s))\sigma dw(s)$$

Now, for a given final time T, using the form of the integral solution $x(t; x_0, u)$ let us introduce operators and sets, which will be used in next parts of the paper [31]. First of all for

$$h_k < T < h_{k+1}, \; k = 0, 1, 2, \ldots, M-1,$$

we define linear and bounded control operator

$$L_T \in L_2^F([0,T], R^m) \to L_2(\Omega, F_T, R^n)$$

as follows

$$L_T u = \sum_{i=0}^{i=k-1} \int_{T-h_{i+1}}^{T-h_i} \left(\sum_{j=0}^{j=i} \exp(A(T-s-h_j))B_j\right) u(s)ds$$
$$+ \int_{0}^{T-h_k} \left(\sum_{j=0}^{j=k} \exp A(T-s-h_j))B_j\right) u(s)ds$$

Moreover, for $T > h_M$ we have

$$L_T u = \sum_{i=0}^{i=M-1} \int_{T-h_{i+1}}^{T-h_i} \left(\sum_{j=0}^{j=i} \exp(A(T-s-h_j))B_j\right) u(s)ds$$
$$+ \int_{0}^{T-h_M} \left(\sum_{j=0}^{j=M} \exp(A(T-s-h_j))B_j\right) u(s)ds$$

11.4 Stationary Systems with Multiple Constant Delays

and its adjoint linear and bounded operator

$$L_T^* \in L_2(\Omega, F_T, R^n) \to L_2^F([0, T], R^m)$$

$$L_T^* z = B_0^* \exp(A^*(T - t)) E\{z|F_t\} \quad \text{for } t \in [0, T - h_M]$$

$$L_T^* z = \left(\sum_{j=1}^{j=i} B_j^* \exp(A^*(T - t - h_j)) \right) E\{z|F_t\}$$
$$\text{for } t \in (T - h_{i+1}, T - h_i], i = 0, 1, \ldots, M - 1$$

From the above notation it follows that the set of all states reachable from initial state $x(0) = x_0 \in L_2(\Omega, F_T, R^n)$ in time $T > 0$, using admissible controls has the form

$$R_T(U_{ad}) = \{x(T; x_0, u) \in L_2(\Omega, F_T, R^n) : u \in U_{ad}\}$$

$$= \exp(At)x_0 + \mathrm{Im} L_T + \int_0^T \exp(A(T - s)) \sigma dw(s)$$

Moreover, we introduce the concept of the linear controllability operator [31, 45, 47, 61, 62] $C_T \in L(L_2(\Omega, F_T, R^n), L_2(\Omega, F_T, R^n))$ which is strongly associated with the control operator L_T and is defined by the following equality

$$C_T = L_T L_T^* = \sum_{i=0}^{i=k-1} \int_{T-h_{i+1}}^{T-h_i} Q(s) Q^*(s) E\{\cdot|F_t\} dt + \int_0^{T-h_k} Q(s) Q^*(s) E\{\cdot|F_t\} dt$$

where

$$Q^*(s) = \sum_{j=0}^{j=i} B_j^* \exp(A^*(T - t - h_j))$$

$$Q(s) = \sum_{j=0}^{j=i} \exp(A(T - t - h_j)) B_j$$

for $h_{i+1} < T < h_i$, $i = 0, 1, \ldots, M - 1$, and for $T > h_M$ by the equality

$$C_T = L_T L_T^* = \sum_{i=0}^{i=M-1} \int_{T-h_{i+1}}^{T-h_i} Q(s) Q^*(s) E\{\cdot|F_t\} dt + \int_0^{T-h_M} Q(s) Q^*(s) E\{\cdot|F_t\} dt$$

Finally, let us recall $n \times n$-dimensional deterministic controllability matrix [31] for $h_{i+1} < T < h_i$, $i = 0, 1,\ldots, M-1$

$$G_T = L_T L_T^* = \sum_{i=0}^{i=k-1} \int_{T-h_{i+1}}^{T-h_i} Q(s)Q^*(s)dt + \int_0^{T-h_k} Q(s)Q^*(s)dt$$

and for $T > h_M$

$$G_T = L_T L_T^* = \sum_{i=0}^{i=M-1} \int_{T-h_{i+1}}^{T-h_i} Q(s)Q^*(s)dt + \int_0^{T-h_M} Q(s)Q^*(s)dt$$

In the proofs of the main results we shall use also controllability operators $C_T(s)$ and controllability matrices $G_T(s)$ depending on time and defining as follows, for $h_{k+1} < T < h_k$, $k = 0, 1,\ldots, M-1$,

$$C_T(s) = L_T L_T^* = \sum_{i=0}^{i=k-1} \int_{T-h_{i+1}}^{T-h_i} Q(s)Q^*(s)E\{\cdot|F_t\}dt + \int_s^{T-h_k} Q(s)Q^*(s)E\{\cdot|F_t\}dt$$

and for $T > h_M$

$$C_T(s) = L_T L_T^* = \sum_{i=0}^{i=M-1} \int_{T-h_{i+1}}^{T-h_i} Q(s)Q^*(s)E\{\cdot|F_t\}dt + \int_s^{T-h_M} Q(s)Q^*(s)E\{\cdot|F_t\}dt$$

Similarly for $h_{k+1} < T < h_k$, $k = 0, 1,\ldots, M-1$, we have

$$G_T(s) = L_T L_T^* = \sum_{i=0}^{i=k-1} \int_{T-h_{i+1}}^{T-h_i} Q(s)Q^*(s)dt + \int_s^{T-h_k} Q(s)Q^*(s)dt$$

and for $T > h_M$

$$G_T(s) L_T L_T^* = \sum_{i=0}^{i=M-1} \int_{T-h_{i+1}}^{T-h_i} Q(s)Q^*(s)dt + \int_s^{T-h_M} Q(s)Q^*(s)dt$$

Remark 11.9 Since the stochastic dynamical system (11.1) is linear, then without loss of generality in the above two definitions it is enough to take zero initial condition $x_0 = 0 \in L_2(\Omega, F_T, R^n)$.

11.4 Stationary Systems with Multiple Constant Delays

Remark 11.10 It should be pointed out, that in the case of delayed controls the above controllability concepts essentially depend on the length of the time interval $[0, T]$.

Remark 11.11 Let us observe, that for the final time $T \leq h_1$ stochastic dynamical system (11.1) is in fact a system without delay.

Remark 11.12 Since the dynamical system (11.3) is stationary, therefore controllability matrix $G_T(s)$ has the same rank at least for all $s \in [0, T - h_k]$, if $h_{k+1} < T < h_k$, $k = 0, 1, \ldots, M - 1$, or for $s \in [0, T - h_M]$, if $T > h_M$, [31].

Remark 11.13 From the form of the controllability operator C_T immediately follows, that this operator is selfadjoint.

In the sequel we study the relationship between the controllability concepts for the stochastic dynamical system (11.4) and controllability of the associated deterministic dynamical system of the form

$$y'(t) = Ay(t) + \sum_{i=0}^{i=M} B_i v(t - h_i) \quad t \in [0, T] \tag{11.6}$$

where the admissible controls $v \in L_2([0, T], R^m)$.

Therefore, let us recall the lemma concerning relative controllability of deterministic system (11.6).

Lemma 11.3 [31]. *The following conditions are equivalent:*

(i) *deterministic system (11.6) is relatively controllable on $[0, T]$,*
(ii) *controllability matrix G_T is nonsingular,*
(iii) rank $[B_0, B_1, \ldots, B_k, \ldots, B_M, AB_0, AB_1, \ldots, AB_M, \ldots, A^{n-1}B_0, A^{n-1}B_1, \ldots, A^{n-1}B_k, \ldots, A^{n-1}B_M] = n$,

Theorem 11.2 *The following conditions are equivalent:*

(i) *Deterministic system (11.6) is relatively controllable on $[0, T]$,*
(ii) *Stochastic system (11.4) is stochastically relatively exactly controllable on $[0, T]$*
(iii) *Stochastic system (11.4) is stochastically relatively approximately controllable on $[0, T]$.*

Proof (i) implies (ii) Let us assume that the deterministic system (11.6) is relatively controllable on $[0, T]$. Then, it is well known (see e.g. [31, 32, 55], or [56]) that the symmetric relative deterministic controllability matrix $G_T(s)$ is invertible and strictly positive definite at least for all $s \in [0, T - h_k]$, if $h_k < T < h_{k+1}$, $k = 0, 1, 2, \ldots, M - 1$ or for at least for all $s \in [0, T - h_M]$ if $T > h_M$, [30]. Hence, for some $\gamma > 0$ we have

$$\langle G_T(s)x, x\rangle \geq \gamma \|x\|^2$$

for all $s \in [0, T - h_M]$ and for all $x \in R^n$.

In order to prove stochastic relative exact controllability on $[0, T]$ for the stochastic system (11.1) we use the relationship between controllability operator C_T and controllability matrix G_T given in Lemma 11.2, to express $E\langle C_T z, z\rangle$ in terms of $\langle G_T Ez, Ez\rangle$. First of all we obtain

$$E\langle C_T z, z\rangle = E\left\langle G_T Ez + \int_0^T G_T(s)q(s)dw(s), Ez + \int_0^T q(s)dw(s) \right\rangle$$

$$= \langle G_T Ez, Ez\rangle$$

$$+ E\int_0^T \langle G_T(s)q(s), q(s)\rangle ds \geq \gamma\left(\|Ez\|^2 + E\int_0^T \|q(s)\|^2 ds\right) = \gamma E\|z\|^2$$

Hence, in the operator sense we have the following inequality $C_T \geq \gamma I$, which means that the operator C_T is strictly positive definite and thus, that the inverse linear operator C_T^{-1} is bounded. Therefore, stochastic relative exact controllability on $[0, T]$ of the stochastic dynamical system (11.1) directly follows from the results given in [31].

(ii) implies (iii) This implication is obvious (see e.g. [15–17, 30]).

(iii) implies (i) Assume that the stochastic dynamical system (11.4) is stochastically relatively approximately controllable on $[0, T]$, and hence its linear self adjoint controllability operator is positive definite, i.e., $C_T > 0$ [8, Chap. 3].

Then, using the resolvent operator $R(\lambda, C_T)$ and following directly the functional analysis method given in [14, 19, 20] for stochastic dynamical systems without delays, we obtain that deterministic system (11.6) is approximately relatively controllable on $[0, T]$.

However, taking into account that the state space for deterministic dynamical system (11.3) is finite dimensional, i.e., exact and approximate controllability coincide [31], we conclude that deterministic dynamical system (11.6) is relatively controllable on $[0, T]$.

Remark 11.14 Let us observe, that for a special case when the final time $T \leq h_1$, stochastic relative exact or approximate controllability problems in $[0, T]$ for stochastic dynamical system with delay in the control (11.1) are reduced to the standard stochastic exact or stochastic approximate controllability problems for the stochastic dynamical system without delays in the control [31].

Corollary 11.2 [14, 19]. *Suppose that $T \leq h_1$. Then the stochastic dynamical control system* (11.4) *is stochastically relatively exactly controllable in $[0, T]$ if and only if*

11.4 Stationary Systems with Multiple Constant Delays

$$\text{rank}\left[B_0, AB_0, A^2 B_0, \ldots, A^{n-1} B_0\right] = n,$$

Corollary 11.3 [19]. *Stochastic dynamical system without delays, i.e. $B_j = 0$, $j = 1, 2, \ldots, M$, is stochastically exactly controllable in any time interval if and only if associated deterministic dynamical system without delay is controllable.*

Remark 11.15 Finally, it should be pointed out, that similarly as for the systems with multiple time variable delays (11.1), using general method given in the monograph [8], for stochastically relatively approximately controllable dynamical systems (11.4) it is possible to formulate the admissible controls $u(t)$, defined for $t \in [t_0, T]$ and transferring given initial state x_0 to the desired final state x_T at time T with minimum energy.

11.5 Systems with Single Time Variable Delay

In the theory of linear, finite-dimensional, time-variable stochastic dynamical control systems we frequently use mathematical model given by the following stochastic ordinary differential state equation with single time-variable point delay in the control

$$dx(t) = (A(t)x(t) + B_0(t)u(t) + B_1(t)u(v(t)))dt + \sigma(t)dw(t) \quad \text{for } t \in [t_0, T], \tag{11.7}$$

with given initial conditions:

$$x(t_0) = x_0 \in L_2(\Omega, F_T, R^n) \quad \text{and } u(t) = 0 \text{ for } t \in [v(t_0), t_0) \tag{11.8}$$

where the state $x(t) \in R^n = X$ and the control $u(t) \in R^m = U$, $A(t) \in L^2([t_0, T], R^{n \times n})$ is $n \times n$ dimensional matrix with square integrable elements, $B_0(t) \in L^2([t_0, T], R^{n \times m})$ and $B_1(t) \in L^2([t_0, T], R^{n \times m})$ are is n × m dimensional matrices with square integrable elements, $s(t) \in L^1([t_0, T], R^{n \times n})$ is $n \times n$ dimensional matrix with continuous elements, $v(t) = t - h(t)$ is continuously differentiable and strictly increasing function, and $h(t) > 0$ is a time-variable point delay [31].

Moreover, since for $[t_0, v(T)]$ dynamical system (1.11) is in fact a system without delay, then we generally assume that the final time $v(T) > t_0$.

In the sequel for simplicity of considerations we introduce the so called leading function $r(t) = t + h(t)$, which is the inverse function for $v(t)$. Therefore, the leading function $r(t) : [v(t_0), v(T)] \to [t_0, T]$ satisfies the following relation $r(v(t)) = t$.

Moreover, it is assumed, that the set of admissible controls is as follows

$$U_{ad} = L_2^F([t_0, T], R^m).$$

It is well known, (see e.g. [31, 32, 61], or [62]) that for a given initial conditions (11.2) and any admissible control $u \in U_{ad}$, for $t \in [v(t_0), t]$ there exist unique solution $x(t; x_0, u) \in L_2(\Omega, F_t, R^n)$ of the linear stochastic differential state Eq. (11.1) which can be represented in the following integral form

$$x(t; x_0, u) = F(t, t_0)x_0 + \int_{t_0}^{t} F(t, s)(B_0(s)u(s) + B_1(s)u(v(s)))ds$$

$$+ \int_{t_0}^{t} F(t, s)\sigma(s)dw(s)$$

where $F(t, s)$ is $n \times n$ dimensional state transition matrix, generated by the matrix $A(t)$.

Thus, taking into account zero initial control for $t \in [v(t_0), t_0)$, the solution for $t \in [t_0, v(T)]$ has the following form [30]

$$x(t; x_0, u) = F(t, t_0)x_0 + \int_{t_0}^{t} F(t, s)B_0(s)u(s)ds + \int_{t_0}^{t} F(t, s)\sigma(s)dw(s)$$

Moreover, for $t > v(T)$ we have

$$x(t; x_0, u) = F(t, t_0)x_0$$

$$+ \int_{t_0}^{t} F(t, s)B_0(s)u(s)ds + \int_{t_0}^{v(t)} F(t, r(s))B_1(r(s))r'(s)u(s)ds$$

$$+ \int_{0}^{t} F(t, s)\sigma(s)dw(s)$$

or changing the order of integration equivalently

$$x(t; x_0, u) = F(t, t_0)x_0$$

$$+ \int_{t_0}^{v(t)} (F(t, s)B_0(s) + F(t, r(s))B_1(r(s))r'(s))u(s)ds$$

$$+ \int_{v(t)}^{t} F(t, s)B_0(s)u(s)ds + \int_{t_0}^{t} F(t, s)\sigma(s)dw(s)$$

11.5 Systems with Single Time Variable Delay

Now, for a fixed given final time T, taking into account the form of the integral solution $x(t; x_0, u)$ let us introduce the following operators and sets [8, Chap. 4].
The linear bounded control operator

$$L_T \in L(L_2^F([0,T], R^m), L_2(\Omega, F_T, R^n))$$

defined by

$$L_T u = \int_{t_0}^{v(T)} (F(t,s)B_0(s) + F(t,r(s))B_1(r(s))r'(s))u(s)ds$$

$$+ \int_{v(T)}^{T} F(t,s)B_0(s)u(s)ds$$

and its adjoint linear bounded operator

$$L_T^* \in L_2(\Omega, F_T, R^n) \to L_2^F([0,T], R^m)$$

$$L_T^* z = (B_0^*(t)F^*(T,t)) + B_1^*(r(t))F^*(T,r(t))r'(t))E\{z|F_t\} \quad \text{for } t \in [t_0, v(T)]$$

$$L_T^* z = B_0^*(t)F^*(T,t)E\{z|F_t\} \quad \text{for } t \in (v(T), T]$$

and the set of all states reachable from initial state

$$x(t_0) = x_0 \in L_2(\Omega, F_T, R^n)$$

in final time T, using admissible controls

$$R_T(U_{ad}) = \{x(T; x_0, u) \in L_2(\Omega, F_T, R^n) : u \in U_{ad}\}$$

$$= F(T, t_0)x_0 + \text{Im}L_T + \int_0^T F(T,s)\sigma(s)dw(s)$$

Moreover, we introduce the concept of the linear controllability operator [31, 61, 62], $C_T \in L(L_2(\Omega, F_T, R^n), L_2(\Omega, F_T, R^n))$, which is strongly associated with the control operator L_T and is defined by the following equality

$$C_T = L_T L_T^*$$

$$= \int_{t_0}^{v(T)} (R_0(t) + r'(t)F(T, r(t))B_1(r(t))B_1^*(r(t))F^*(T, r(t))r'(t)E\{\cdot|F_t\}dt$$

$$+ \int_{v(T)}^{T} R_0(t)E\{\cdot|F_t\}dt$$

where

$$R_0(t) = F(T,t)B_0(t)B_0^*(t)F^*(T,t)$$

Finally, let us recall $n \times n$-dimensional deterministic controllability matrix [31, 61]

$$G_T = \int_{t_0}^{v(T)} (R_0(t) + r'(t)F(T,r(t))B_1(r(t))B_1^*(r(t))F^*(T,r(t))r'(t))dt$$

$$+ \int_{v(T)}^{T} R_0(t)dt$$

In the proofs of the main results we shall use also controllability operators $C_T(s)$ and controllability matrices $G_T(s)$ depending on time $s \in [t_0, v(T)]$, and defining as follows,

$$C_T(s) = L_T(s)L_T^*(s)$$

$$= \int_{s}^{v(T)} (R_0(t) + r'(t)F(T,r(t))B_1(r(t))B_1^*(r(t))F^*(T,r(t))r'(t))E\{\cdot|F_t\}dt$$

$$+ \int_{v(T)}^{T} R_0(t)E\{\cdot|F_t\}dt$$

$$G_T(s) = \int_{s}^{v(T)} (R_0(t) + r'(t)F(T,r(t))B_1(r(t))B_1^*(r(t))F^*(T,r(t))r'(t))dt$$

$$+ \int_{v(T)}^{T} R_0(t)dt$$

Remark 11.16 Let us recall once again, that, since for the final time T such that $v(T) \leq t_0$ stochastic dynamical system (11.7) is in fact a system without delay in the control therefore, in the sequel we generally assumed that $v(T) > t_0$.

Remark 11.17 From the form of the controllability operator C_T immediately follows, that this operator is self adjoint.

In the sequel we study the relationship between the controllability concepts for the stochastic dynamical system (11.7) and controllability of the associated deterministic time-variable finite-dimensional dynamical system with single time-varying point delay in the control of the following form

11.5 Systems with Single Time Variable Delay

$$y'(t) = A(t)y(t) + B_0(t)w(t) + B_1(t)w(v(t)) \qquad t \in [t_0, T] \qquad (11.9)$$

where the admissible controls $w \in L_2([t_0, T], R^m)$.

Theorem 11.3 *The following conditions are equivalent:*

(i) *Deterministic system (11.9) is relatively controllable on $[t_0, T]$,*
(ii) *Stochastic system (11.7) is stochastically relatively exactly controllable on $[t_0, T]$*
(iii) *Stochastic system (11.7) is stochastically relatively approximately controllable on $[t_0, T]$.*

Proof (i) implies (ii) Let us assume that the deterministic system (11.9) is relatively controllable on $[t_0, T]$. Then, it is well known, (see e.g. [31], [12]) that the relative deterministic controllability matrix $G_T(s)$ is invertible and strictly positive definite at least for all $s \in [t_0, v(T)]$, [31]. Hence, for some $\gamma > 0$ we have

$$\langle G_T(s)x, x \rangle \geq \gamma \|x\|^2$$

for all $s \in [t_0, v(T)]$ and for all $x \in R^n$. In order to prove stochastic relative exact controllability on $[t_0, v(T)]$ for the stochastic system (11.1) we use the relationship between controllability operator C_T and controllability matrix G_T given in Lemma 11.2, to express $E\langle C_T z, z \rangle$ in terms of $\langle G_T Ez, Ez \rangle$.

First of all using formulas for relative controllability operator C_T and relative controllability matrix G_T we obtain

$$E\langle C_T z, z \rangle = E\left\langle G_T Ez + \int_{t_0}^{T} G_T(s)q(s)dw(s), Ez + \int_{t_0}^{T} q(s)dw(s) \right\rangle$$

$$= \langle G_T Ez, Ez \rangle$$

$$+ E\int_{t_0}^{T} \langle G_T(s)q(s), q(s) \rangle ds \geq \gamma \left(\|Ez\|^2 + E\int_{t_0}^{T} \|q(s)\|^2 ds \right) = \gamma E\|z\|^2$$

Hence, in the operator sense we have the following inequality $C_T \geq \gamma I$, which means that the operator C_T is strictly positive definite and thus, that the inverse linear operator C_T^{-1} is bounded.

Therefore, stochastic relative exact controllability on $[t_0, T]$ of the stochastic dynamical system (11.3) directly follows from the results given in [31].

(ii) implies (iii) This implication is obvious (see e.g. [31, 61, 62]).

(iii) $\Rightarrow$ (i) Assume that the stochastic dynamical system (11.7) is stochastically relatively approximately controllable on $[t_0, T]$, and hence its controllability operator is positive definite, i.e., $C_T > 0$ [31].

Then, using the resolvent operator $R(\lambda, C_T)$ and following directly the functional analysis method given in [55, 56, 61, 62] for stochastic dynamical systems without delays, we obtain that deterministic system (11.3) is approximately relatively controllable on $[t_0, T]$.

However, taking into account that the state space for deterministic dynamical system (11.6) is finite dimensional, i.e., exact and approximate controllability coincide [31], we conclude that deterministic dynamical system (11.6) is relatively controllable on $[t_0, T]$.

Remark 11.18 Let us observe, that for a special case, when the final time is such that $v(T) \leq t_0$, stochastic relative exact or approximate controllability problems in $[t_0, T]$ for stochastic dynamical system with delay in the control (11.4) are reduced to the standard stochastic exact or stochastic approximate controllability problems for the stochastic dynamical system without delays in the control [31].

Finally, we shall consider stationary dynamical system with constant matrices A, B_0, B_1 and single constant point delay $h > 0$ in the control. In this very special case, without loss of generality we can take the initial time $t_0 = 0$. Then, taking into account Theorem 11.2 or Theorem 11.3 and relative controllability conditions for stationary deterministic dynamical systems with single constant point delay given in [31] we can formulate the following two simple corollaries.

Corollary 11.4 *Suppose that $h < T$. Then stationary stochastic dynamical control system (11.7) is stochastically relatively exactly controllable in $[0, T]$ if and only if*

$$\operatorname{rank}\left[B_0, B_1, AB_0, AB_1, A^2B_0, A^2B_1, \ldots, A^{n-1}B_0, A^{n-1}B_1\right] = n,$$

Corollary 11.5 [14, 19]. *Suppose that $T \leq h$. Then stationary stochastic dynamical control system (11.7) is stochastically relatively exactly controllable in $[0, T]$ if and only if*

$$\operatorname{rank}\left[B_0, AB_0, A^2B_0, \ldots, A^{n-1}B_0\right] = n,$$

Remark 11.19 Finally, it should be pointed out, that similarly as for the systems with multiple delays (11.1), using very general method given in the monograph [31], it is possible to formulate for stochastically relatively approximately controllable dynamical systems (11.7) the analytic formula for the admissible controls $u(t)$, defined for $t \in [t_0, T]$ and transferring given initial state x_0 to the desired final state x_T at time T with minimum energy.

11.6 Minimum Energy Control

Minimum energy control problem is strongly connected with controllability concept, (see e.g., [31] for more details). First of all, let us observe, that for exactly controllable on $[t_0, T]$ linear control system there exists generally many different admissible controls $u(t)$, defined for $t \in [t_0, T]$ and transferring given initial state x_0 to the desired final state x_T at time T.

11.6 Minimum Energy Control

Therefore, we may ask which of these possible admissible controls are optimal one according to given a priori criterion. In the sequel we shall consider minimum energy control problem for stochastic dynamical system (11.1) with the optimality criterion representing the energy of control. In this case optimality criterion has the following simple form

$$J(u) = E \int_{t_0}^{T} \|u(t)\|^2 dt$$

In order to simplify the considerations let us assume that the final time $T > r_M(t_0)$ (see Remark 11.9), and let us introduce the following notations:

$$B_T(t) = \sum_{j=0}^{j=i} F(t_0, r_j(t)) B_j(r_j(t)) r'_j(t)$$

for $t \in [r_i(t_0), r_{i+1}(t_0)]$, and $i = 0, 1, 2, \ldots, M-1$

and

$$B_T(t) = \sum_{j=0}^{j=M} F(t_0, r_j(t)) B_j(r_j(t)) r'_j(t) \quad \text{for } t \in (r_M(t_0), T]$$

Theorem 11.4 *Assume that the stochastic dynamical system (11.4) is stochastically relatively exactly controllable on [0, T]. Then for arbitrary $x(t_0) = x_0 \in L_2(\Omega, F_T, R^n)$ and arbitrary σ, the control*

$$u^0(t) = B_T^*(t) F^*(T, t)) E \left\{ C_T^{-1} \left(x_T - F(T, t_0) x_0 - \int_{t_0}^{T} F(T, s) \sigma(s) dw(s) \right) | F_t \right\}$$

for $t \in [t_0, T]$ transfers the system (11.1) from initial state x_0 to the final state x_T at time T.

Moreover, among all admissible controls $u^a(t)$ transferring initial state x_0 to the final state x_T at time T, the control $u^0(t)$ minimizes the integral performance index

$$J(u) = E \int_{t_0}^{T} \|u(t)\|^2 dt$$

Proof First of all let us observe, that since the stochastic dynamical system (11.1) is stochastically relatively exactly controllable on $[t_0, T]$, then the controllability operator C_T is invertible and its inverse C_T^{-1} is a linear and bounded operator, i.e.,

$$C_T^{-1} \in L(L_2(\Omega, F_T, R^n), L_2(\Omega, F_T, R^n)).$$

Substituting the control $u^0(t)$ into the solution formula of the differential state equation, one can easily obtain

$$x(t; x(0), u^0(t)) = F(t, t_0)x_0$$
$$+ \int_{t_0}^{t} F(t,s)B_T(t)B_T^*(t)F^*(t,s))E$$
$$\left\{ C_T^{-1}\left(x_T - F(T, t_0)x_0 - \int_{t_0}^{T} F(T,s)\sigma(s)dw(s) \right) \right\} | F_s ds$$
$$+ \int_{t_0}^{t} F(t,s)\sigma(s)dw(s)$$

for $t \in [0, T]$

Hence, for $t = T$ we have,

$$x(T; x_0, u^0(t)) = F(T, t_0)x_0$$
$$+ \int_{t_0}^{T} \left(F(T,s))B_T(t)B_T^*(t)F^*(T,s)) \right)E$$
$$\left\{ C_T^{-1}\left(x_T - F(T, t_0)x_0 - \int_{t_0}^{T} F(T,s)\sigma(s)dw(s) \right) \right\} | F_s ds$$
$$+ \int_{t_0}^{T} F(T,s)\sigma(s)dw(s)$$

Thus, taking into account the form of the operator C_T we have

$$x(T; x_0, u^0(t)) = F(T, t_0)x_0 + C_T C_T^{-1}\left(x_T - \exp(AT)x(0) - \int_{t_0}^{T} \exp(A(T-s))\sigma dw(s) \right)$$
$$+ \int_{t_0}^{T} \exp(A(T-s))\sigma dw(s)$$
$$= F(T, t_0)x_0 + x_T - F(T, t_0)x_0$$
$$- \int_{t_0}^{T} F(T,s)\sigma(s)dw(s) + \int_{t_0}^{T} F(T,s)\sigma(s)dw(s) = x_T$$

11.6 Minimum Energy Control

Therefore for $t = T$ we see that the control $u^0(t)$ transfers dynamical system (11.1) from given initial state $x_0 \in L_2(\Omega, F_T, R^n)$ to the desired final state

$$x_T \in L_2(\Omega, F_T, R^n)$$

at time T.

In the second part of the proof using the method presented in [31] we shall show that the control $u^0(t)$, $t \in [t_0, T]$ is optimal according to performance index J. In order to do that, let us suppose that $u'(t)$, $t \in [t_0, T]$ is any other admissible control which also steers the initial state x_0 to the final state x_T at time T. Hence using controllability operator defined in Sect. 11.2 we have

$$L_T(u^0(\cdot)) = L_T(u'(\cdot))$$

Subtracting from both sides and using the properties of scalar product in the space R^n and the form of controllability operator L_T we obtain the following equality

$$E \int_{t_0}^{T} \langle (u'(t) - u^0(t)), u^0(t) \rangle dt = 0$$

Moreover, using once again properties of the scalar product in R^n we have

$$E \int_{t_0}^{T} \|u'(t)\|^2 dt = E \int_{t_0}^{T} \|u'(t) - u^0(t)\|^2 dt + E \int_{t_0}^{T} \|u^0(t)\|^2 dt$$

$$\text{Since } E \int_{t_0}^{T} \|u'(t) - u^0(t)\|^2 dt \geq 0,$$

we conclude, that for any admissible control $u'(t)$, $t \in [t_0, T]$ the following inequality holds

$$E \int_{t_0}^{T} \|u^0(t)\|^2 dt \leq E, \int_{t_0}^{T} \|u'(t)\|^2 dt$$

Hence the control $u^0(t)$, $t \in [t_0, T]$ is optimal control according to the performance index J, and thus it is minimum energy control.

Remark 11.20 Let us observe, that for the case when the final time $T \leq r_1(t_0)$, minimum energy control problem in $[t_0, T]$ for stochastic dynamical system with delays in the control (11.1) is in fact reduced to the minimum energy control problem for the stochastic dynamical system without delays in the control. Therefore, in this case the solution of minimum energy control problem is given by Theorem 11.4 with the substitution $M = 0$.

11.7 Minimum Energy Control of Stationary Systems with Multiple Delays

In the sequel we shall consider minimum energy control problem for stationary stochastic dynamical system (11.4) with the optimality criterion representing the energy of control. In this special case optimality criterion has the following simple form

$$J(u) = E \int_{t_0}^{T} \|u(t)\|^2 dt$$

In order to simplify the considerations let us assume that the final time $T > h_M$ (see Remark 11.9), and let us introduce the following notations:

$$B_T(t) = \sum_{j=0}^{j=i} \exp(A(t+h_j))B_j \quad \text{for } t \in [h_i, h_{i+1}], \quad i = 0, 1, 2, \ldots, M-1$$

and

$$B_T(t) = \sum_{j=0}^{j=M} \exp(A(t+h_j))B_j \quad \text{for } t \in [h_M, T]$$

Theorem 11.5 *Assume that the stochastic dynamical system (11.4) is stochastically relatively exactly controllable on $[0, T]$. Then for arbitrary $x_T \in L_2(\Omega, F_T, R^n)$ and arbitrary σ, the control*

$$R_T(U_{ad}) = \{x(T; x_0, u) \in L_2(\Omega, F_T, R^n) : u \in U_{ad}\}$$

$$= F(T, t_0)x_0 + \text{Im } L_T + \int_{t_0}^{T} F(T, s))\sigma(s)dw(s) \quad \text{for } t \in [0, T]$$

transfers the system (11.4) from initial state x_0 to the final state x_T at time T. Moreover, among all admissible controls $u'(t)$ transferring initial state x_0 to the final state x_T at time T, the control $u^0(t)$ minimizes the integral performance index

11.7 Minimum Energy Control of Stationary Systems with Multiple Delays

$$J(u) = E \int_0^T \|u(t)\|^2 dt$$

Proof First of all let us observe, that since the stochastic dynamical system (11.4) is stochastically relatively exactly controllable on $[0, T]$, then the controllability operator C_T is invertible and its inverse C_T^{-1} is a linear and bounded operator, i.e.,

$C_T = L_T L_T^*$

$$= \sum_{i=0}^{i=k-1} \int_{v_{i+1}(T)}^{v_i(T)} \left(\sum_{j=0}^{j=i} F(T, r_j(s)) B_j(r_j(s)) r'_j(s) \right) \left(\sum_{j=0}^{j=i} B_j^*(r_j(s)) F^*(T, r_j(s)) r'_j(s) \right) E\{\cdot | F_t\} dt$$

$$+ \int_{t_0}^{v_k(T)} \left(\sum_{j=0}^{j=i} F(T, r_j(s)) B_j(r_j(s)) r'_j(s) \right) \left(\sum_{j=0}^{j=i} B_j^*(r_j(s)) F^*(T, r_j(s)) r'_j(s) \right) E\{\cdot | F_t\} dt$$

Moreover,

$G_T = L_T L_T^*$

$$= \sum_{i=0}^{i=M-1} \int_{v_{i+1}(T)}^{v_i(T)} \left(\sum_{j=0}^{j=i} F(T, r_j(s)) B_j(r_j(s)) r'_j(s) \right) \left(\sum_{j=0}^{j=i} B_j^*(r_j(s)) F^*(T, r_j(s)) r'_j(s) \right) dt$$

$$+ \int_{t_0}^{v_M(T)} \left(\sum_{j=0}^{j=M} F(T, r_j(s)) B_j(r_j(s)) r'_j(s) \right) \left(\sum_{j=0}^{j=M} B_j^*(r_j(s)) F^*(T, r_j(s)) r'_j(s) \right) dt$$

Hence, substituting the control $u^0(t)$ into the solution formula of the differential state equation, and taking into account the form of the operator C_T and G_T one can easily obtain, that for $t = T$ the control $u^0(t)$ transfers dynamical system (11.1) from given initial state $x_0 \in L_2(\Omega, F_T, R^n)$ to the desired final state $x_T \in L_2(\Omega, F_T, R^n)$ at time T.

In the second part of the proof using the method presented in [31] we shall show that the control $u^0(t)$, $t \in [0, T]$ is optimal according to performance index J. In order to do that, let us suppose that $u'(t)$, $t \in [0, T]$ is any other admissible control, which also steers the initial state x_0 to the final state x_T at time T. Hence using controllability operator defined in Sect. 11.2 we have

$$L_T(u^0(\cdot)) = L_T(u'(\cdot))$$

Subtracting from both sides and using the properties of scalar product in the space R^n and the form of controllability operator L_T we obtain the following equality

$$E \int_{t_0}^{T} \langle (u'(t) - u^0(t)), u^0(t) \rangle dt = 0$$

Moreover, using once again properties of the scalar product in R^n we have

$$E \int_{t_0}^{T} \|u'(t)\|^2 dt = E \int_{t_0}^{T} \|u'(t) - u^0(t)\|^2 dt + E \int_{t_0}^{T} \|u^0(t)\|^2 dt$$

Since

$$E \int_{t_0}^{T} \|u'(t) - u^0(t)\|^2 dt \geq 0,$$

we conclude that for any admissible control $u'(t)$, $t \in [0, T]$ the following inequality holds

$$E \int_{t_0}^{T} \|u^0(t)\|^2 dt \leq E \int_{t_0}^{T} \|u'(t)\|^2 dt$$

Hence the control $u^0(t)$, $t \in [0, T]$ is optimal control according to the performance index J, and thus it is minimum energy control.

Remark 11.21 Let us observe, that for the case when the final time $T \in (h_i, h_{i+1}]$, minimum energy control problem in $[0, T]$ for stochastic dynamical system with delays in the control (11.1) is in fact reduced to the minimum energy control problem for the stochastic dynamical system i delays in the control. Therefore, in this case the solution of minimum energy control problem is given by Theorem 11.2 with the substitution $M = i$.

11.8 Minimum Energy Control of Systems with Single Delay

In the sequel we shall consider minimum energy control problem for nonstationary stochastic dynamical system (11.1) with single time-variable delay in the control and with the optimality criterion representing the energy of control. In this case optimality criterion has the following simple form

11.8 Minimum Energy Control of Systems with Single Delay

$$J(u) = E \int_{t_0}^{T} \|u(t)\|^2 dt$$

Theorem 11.6 *Assume that the stochastic dynamical system (11.1) is relatively exactly controllable on $[t_0, T]$. Then for arbitrary $x_T \in L_2(\Omega, F_T, R^n)$ and arbitrary σ, the control*

$$u^0(t) = B_0^*(t) F^*(T,t) E \left\{ C_T^{-1} \left(x_T - F(T,t_0) x_0 - \int_{t_0}^{v(T)} F(T,s)\sigma(s) dw(s) \right) | F_t \right\}$$

for $t \in [t_0, v(T)]$

$$u^0(t) = \left(B_0^*(t) F^*(T,t) + r'(t) B_1^*(t) F^*(T, r(t)) \right)$$
$$\times E \left\{ C_T^{-1} \left(x_T - F(T,t_0) x_0 - \int_{v(T)}^{T} F(T,s)\sigma(s) dw(s) \right) | F_t \right\}$$

for $t \in [v(T), T]$
transfers the system (11.1) from initial state x_0 to the final state x_T at time $T > v(T)$.

Moreover, among all admissible controls $u'(t)$ transferring initial state x_0 to the final state x_T at time $T > v(T)$, the control $u^0(t)$ minimizes the integral performance index

$$J(u) = E \int_{t_0}^{T} \|u(t)\|^2 dt$$

Proof First of all let us observe, that since the stochastic dynamical system (11.1) is stochastically relatively exactly controllable on $[t_0, T]$, then the controllability operator C_T is invertible and its inverse C_T^{-1} is a linear and bounded operator, i.e.,

$$C_T^{-1} \in L(L_2(\Omega, F_T, R^n), L_2(\Omega, F_T, R^n)).$$

Substituting the control $u^0(t)$ into the solution formula of the differential state equation, one can easily obtain

$$x(t;x_0,u^0(t)) = F(t,t_0)x_0$$
$$+ \int_0^t F(t,s)B_0(s)B_0^*(s)F^*(t,s)E$$
$$\left\{ C_T^{-1}\left(x_T - F(T,s)x_0 - \int_{t_0}^{v(T)} F(T,s)\sigma(s)dw(s) \right) \right\} |F_s ds$$
$$+ \int_{t_0}^t F(T,s)\sigma(s)dw(s)$$

for $t \in (t_0, v(T)]$

$$x(t;x_0,u^0(t)) = F(t,t_0)x_0$$
$$+ \int_0^{v(t)} (F(t,s)B_0(s)B_0^*(s)F^*(T,s)$$
$$+ r'(s)F(T,r(s))B_1(r(s))B_1^*(r(s))F^*(T,r(s))r'(s))$$
$$\times E\left\{ C_T^{-1}\left(x_T - F(T,s)x_0 - \int_{v(T)}^T F(T,s)\sigma(s)dw(s) \right) \right\} |F_s ds$$
$$+ \int_{v(t)}^t (F(t,s)B_0(s)B_0^*(s)F^*(T,s)E$$
$$\left\{ C_T^{-1}\left(x_T - F(T,t_0)x_0 - \int_{t_0}^{v(T)} F(T,s)\sigma(s)dw(s) \right) \right\} |F_s ds$$
$$+ \int_{t_0}^t F(t,s)\sigma(s)dw(s)$$

for $t \in (v(T), T]$

Hence, for $t = T$ we have

11.8 Minimum Energy Control of Systems with Single Delay

$$x(T;x_0,u^0(t)) = F(T,t_0)x_0$$

$$+ \left(\int_{t_0}^{v(T)} F(T,s)B_0(s)B_0^*(s)F^*(T,s) + r'(s)F(T,r(s))B_1(s)B_1^*(s)F^*(T,r(s))r'(s) \right)$$

$$\times E\left\{ C_T^{-1}\left(x_T - F(T,t_0)x_0 - \int_{v(T)}^{T} F(T,s)\sigma(s)dw(s) \right) \right\} |F_s ds$$

$$+ \int_{v(t)}^{T} F(T,s)B_0(s)B_0^*(s)F^*(T,s)E$$

$$\left\{ C_T^{-1}\left(x_T - F(T,s)x_0 - \int_{t_0}^{v(T)} F(T,s)\sigma(s)dw(s) \right) \right\} |F_s ds$$

$$+ \int_{t_0}^{T} F(T,s)\sigma(s)dw(s)$$

Thus, taking into account the form of the operator C_T we have

$$x(T;x_0,u^0(t)) = F(T,t_0)x_0 + C_T C_T^{-1}\left(x_T - F(T,t_0)x_0 - \int_{t_0}^{t} F(T,s)\sigma(s)dw(s) \right)$$

$$+ \int_{t_0}^{T} F(t,s)\sigma(s)dw(s)$$

$$= F(T,t_0)x_0 + x_T - F(T,t_0)x_0$$

$$- \int_{t_0}^{T} F(T,s)\sigma(s)dw(s) + \int_{t_0}^{T} F(T,s)\sigma(s)dw(s)$$

$$= x_T$$

Therefore, for $t = T$ we see that the control $u^0(t)$ transfers dynamical system (1) from given initial state $x_0 \in L_2(\Omega, F_T, R^n)$ to the desired final state $x_T \in L_2(\Omega, F_T, R^n)$ at time T > v(T).

In the second part of the proof using the method presented in [31] we shall show that the control $u^0(t)$, $t \in [t_0,T]$ is optimal according to performance index J. In order to do that, let us suppose that $u'(t)$, $t \in [t_0,T]$ is any other admissible control which also steers the initial state x_0 to the final state x_T at time T. Hence, using controllability operator defined in Sect. 11.2 we have

$$L_T(u^0(\cdot)) = L_T(u'(\cdot))$$

Subtracting from both sides and using the properties of scalar product in the space R^n and the form of controllability operator L_T we obtain the following equality

$$E \int_{t_0}^{T} \langle (u'(t) - u^0(t)), u^0(t) \rangle dt = 0$$

Moreover, using once again properties of the scalar product in R^n we have

$$E \int_{t_0}^{T} \|u'(t)\|^2 dt = E \int_{t_0}^{T} \|u'(t) - u^0(t)\|^2 dt + E \int_{t_0}^{T} \|u^0(t)\|^2 dt$$

Since

$$E \int_{t_0}^{T} \|u'(t) - u^0(t)\|^2 dt \geq 0,$$

we conclude that for any admissible control $u'(t)$, $t \in [0, T]$ the following inequality holds

$$E \int_{t_0}^{T} \|u^0(t)\|^2 dt \leq E \int_{t_0}^{T} \|u'(t)\|^2 dt$$

Hence the control $u^0(t)$, $t \in [t_0, T]$ is optimal control according to the performance index J, and thus it is minimum energy control.

In the sequel, as special case we shall consider minimum energy control problem for stationary stochastic dynamical system with single constant point delay in the control and with the optimality criterion representing the energy of control. In this case optimality criterion has the following simple form

$$J(u) = E \int_{0}^{T} \|u(t)\|^2 dt$$

Theorem 11.7 *Assume that the stochastic dynamical system (11.1) is relatively exactly controllable on $[0, T]$. Then, for arbitrary $x_T \in L_2(\Omega, F_T, R^n)$ and arbitrary σ, the control*

11.8 Minimum Energy Control of Systems with Single Delay

$$u^0(t) = B_0^* \exp(A^*(T-t)) E\left\{ C_T^{-1}\left(x_T - \exp(AT)x_0 - \int_0^{T-h} \exp(A(T-s))\sigma dw(s) \right) | F_t \right\}$$

for $t \in [0, h]$

$$u^0(t) = \left(B_0^* \exp(A^*(T-t)) + B_1^* \exp(A^*(T-h-t)) \right)$$
$$\times E\left\{ C_T^{-1}\left(x_T - \exp(AT)x_0 - \int_{T-h}^T \exp(A(T-h-s))\sigma dw(s) \right) | F_t \right\}$$

for $t \in [h, T]$ transfers the system (11.1) from initial state x_0 to the final state x_T at time $T > h$.

Moreover, among all admissible controls $u^a(t)$ transferring initial state x_0 to the final state x_T at time $T > h$, the control $u^0(t)$ minimizes the integral performance index

$$J(u) = E \int_0^T \|u(t)\|^2 dt$$

Proof First of all let us observe, that since the stochastic dynamical system (11.1) is stochastically relatively exactly controllable on $[0, T]$, then the controllability operator C_T is invertible and its inverse C_T^{-1} is a linear and bounded operator, i.e.,

$$C_T^{-1} \in L(L_2(\Omega, F_T, R^n), L_2(\Omega, F_T, R^n)).$$

Substituting the control $u^0(t)$ into the solution formula of the differential state equation, and taking into account the form of the operators C_T and G_T one can easily obtain, that for $t = T$ we have.

Thus, taking into account the form of the operator C_T and G_T, we see that for $t = T$ the control $u^0(t)$ transfers dynamical system (11.1) from given initial state $x_0 \in L_2(\Omega, F_T, R^n)$ to the desired final state $x_T \in L_2(\Omega, F_T, R^n)$ at time $T > h$.

In the second part of the proof using the method presented in [8, Chap. 1] we shall show that the control $u^0(t)$, $t \in [0, T]$ is optimal according to performance index J. In order to do that, let us suppose that $u'(t)$, $t \in [0, T]$ is any other admissible control which also steers the initial state x_0 to the final state x_T at time T. Hence using controllability operator defined in Sect. 11.2 we have

$$L_T(u^0(\cdot)) = L_T(u'(\cdot))$$

Subtracting from both sides and using the properties of scalar product in the space R^n and the form of controllability operator L_T we obtain the following equality

$$E \int_{t_0}^{T} \langle (u'(t) - u^0(t)), u^0(t) \rangle dt = 0$$

Moreover, using once again properties of the scalar product in R^n we have

$$E \int_{t_0}^{T} \|u'(t)\|^2 dt = E \int_{t_0}^{T} \|u'(t) - u^0(t)\|^2 dt + E \int_{t_0}^{T} \|u^0(t)\|^2 dt$$

Since

$$E \int_{t_0}^{T} \|u'(t) - u^0(t)\|^2 dt \geq 0,$$

we conclude that for any admissible control $u'(t)$, $t \in [0, T]$ the following inequality holds

$$E \int_{t_0}^{T} \|u^0(t)\|^2 dt \leq E \int_{t_0}^{T} \|u'(t)\|^2 dt$$

Hence the control $u^0(t)$, $t \in [0, T]$ is optimal control according to the performance index J, and thus it is minimum energy control.

In this Chapter sufficient conditions for stochastic relative exact controllability for linear stationary finite-dimensional stochastic control systems with multiple both time-variable and constant point delays in the control have been formulated and proved, respectively.

These conditions extend to the case of one constant point delay in control, known stochastic exact controllability conditions for dynamical control systems without delays recently published in the papers [55, 56, 61, 62].

Finally, it should be pointed out, that using the standard techniques presented in the monograph [31] it is possible to extend the results presented in this paper for more general nonstationary linear stochastic control systems with many time variable point delays in the control.

Moreover, the extension for stochastic absolute exact controllability and stochastic absolute approximate controllability in a given time interval is also possible. Finally, using methods taken directly from functional analysis minimum energy control problems for all types of stochastic systems is formulated and analytically solved.

Chapter 12
Controllability of Stochastic Systems with Distributed Delays in Control

12.1 Introduction

In the present Chapter we shall study stochastic controllability problems for linear dynamical systems, which are natural generalizations of controllability concepts well known in the theory of infinite dimensional control systems [30, 31]. More precisely, we shall consider stochastic relative exact and approximate controllability problems for finite-dimensional linear stationary dynamical systems with multiple constant point delays in the control described by stochastic ordinary differential state equations.

More precisely, using techniques similar to those presented in the papers [38, 39, 45, 47, 61, 62] we shall formulate and prove necessary and sufficient conditions for stochastic relative exact controllability in a prescribed time interval for linear stationary stochastic dynamical systems with multiple constant point delays in the control.

Roughly speaking, it will be proved that under suitable assumptions relative controllability of a deterministic linear associated dynamical system is equivalent to stochastic relative exact controllability and stochastic relative approximate controllability of the original linear stochastic dynamical system. This is a generalization to control delayed case some previous results concerning stochastic controllability of linear dynamical systems without delays in control [61, 62].

The present Chapter is organized as follows: Sect. 12.2 contains general mathematical model of linear, finite-dimensional stationary stochastic dynamical system with distributed delays in the control. Moreover, in this section basic notations and definitions of stochastic relative exact and stochastic approximate controllability are presented. It should be pointed out, that for selfcontainess some preliminary results concerning stochastic processes are also included.

In Sect. 12.3 using results and methods taken directly from deterministic controllability problems, necessary and sufficient conditions for exact and approximate stochastic relative controllability are formulated and proved. Finally, concluding remarks and states some open controllability problems for more general stochastic dynamical systems are presented at the end of Chapter.

12.2 System Description

Throughout this paper, unless otherwise specified, we use the following standard notations. Let (Ω, F, P) be a complete probability space with probability measure P on Ω and a filtration $\{F_t | t \in [0, T]\}$ generated bmy n-dimensional Wiener process $\{w(s): 0 \leq s \leq t\}$ defined on the probability space (Ω, F, P).

Let $L_2(\Omega, F_T, R^n)$ denote the Hilbert space of all F_T-measurable square integrable random variables with values in R^n. Moreover, let $L_2^F([0, T], R^n)$ be the Hilbert space of all square integrable and F_t-measurable random variables with values in finite dimensional space R^n.

Let $h > 0$ be given. For a function $u: [t_0-h, t_1] \to R^m$ and for $t \in [t_0, t_1]$, we use the symbol u_t to denote the function on $[-h, 0)$ defined by $u_t(s) = u(t+s)$ for $s \in [-h, 0)$. Let the linear stochastic control system with distributed delay in control be modelled by the Ito equation of the following form:

$$dx(t) = \left[A(t)x(t) + \int_{-h}^{0} u(t+s) dH_s(t,s) \right] dt + \sigma(t) d\omega(t) \quad (12.1)$$

$$x(t_0) = x_0$$

where $x(t) \in R^n$ is the vector describing the instantaneous state of the stochastic system and $\sigma : [t_0, t_1] \times R^n \to R^{n \times n}$.

The continuous matrix $A(t) \in R^{n \times n}$ is the state matrix, $u(t) \in R^m$ is a vector input to the stochastic dynamical system, $H(t, s)$ is $n \times m$-dimensional continuous matrix in t for fixed s and is of bounded variation in s on the interval $[-h, 0]$ and ω is n-dimensional Wiener process. The integral is in the Lebesque-Stieltjes sense with respect to the variable s.

By using the well known method and un-symmetric Fubini theorem we obtain the explicit solution of the delayed control system for $t > 0$ in the following compact form:

$$x(t) = \Phi(t, t_0) x_0 + \int_{t_0}^{t} \Phi(t, \tau) \left(\int_{-h}^{0} u(\tau + s) d_s H(\tau, s) \right) d\tau + \int_{t_0}^{t} \Phi(t, s) \sigma(s) d\omega(s)$$

$$= \Phi(t, t_0) x_0 + \int_{-h}^{0} d_H \left(\int_{t_0 + s}^{t_0} \Phi(t, \tau - s) H(\tau - s, s) u_{t_0} d\tau \right)$$

$$+ \int_{t_0}^{t} \left(\int_{-h}^{0} \Phi(t, \tau - s) d H_t(\tau - s, s) \right) u(\tau) d\tau + \int_{t_0}^{t} \Phi(t, s) \sigma(s) d\omega(s)$$

12.2 System Description

where $\Phi(t, t_0)$ is the fundamental matrix of the homogeneous equation $\dot{x}(t) = A(t)x(t)$ with initial condition $x(t_0) = x_0$, symbol d_H denotes that the integration is in the Lebesque-Stieltjes sense with respect to the variable s in matrix H and

$$H_t(\tau, s) = \begin{cases} H(\tau, s) & \text{for } \tau \leq t \\ 0 & \text{for } \tau > t \end{cases}$$

In the sequel, for simplicity of considerations, we shall assume that the set of admissible controls is $U_{ad} = L_2^F([t_0, t_1], R^m)$ and let

$$S(\tau, t_1) = \int_{-h}^{0} \Phi(t_1, \tau - s) dH_{t_1}(\tau - s, s).$$

Now let us introduce the following matrices and sets:
The linear bounded operator $L: U_{ad} \to L_2(\Omega, F_{t_1}, R^n)$ is defined by

$$Lu = \int_{t_0}^{t_1} S(s, t_1) u(s) ds$$

and its adjoint linear bounded operator

$$L^*: L_2(\Omega, F_{t_1}, R^n) \to U_{ad}$$

is defined by

$$(L^*z)(t) = S^*(t, t_1) E\{z|F_t\}, \quad t \in [t_0, t_1]$$

and the attainable set of all states attainable from x_0 in time $t > 0$ using admissible controls is defined by

$$R_t(U_{ad}) = \{x(t; x_0, u) \in L_2(\Omega, F_{t_1}, R^n) : u(\cdot) \in U_{ad}\}$$

$$= \Phi(t, t_0) x_0 + \int_{-h}^{0} d_H \left(\int_{t_0 + s}^{t_0} \Phi(t, \tau - s) H(\tau - s, s) u_{t_0} d\tau \right)$$

$$+ \operatorname{Im} L + \int_{t_0}^{t} \Phi(t, s) \sigma(s) d\omega(s)$$

The linear bounded controllability operator

$$W : L_2(\Omega, F_{t_1}, R^n) \rightarrow L_2(\Omega, F_{t_1}, R^n)$$

which is associated with the operator L is defined by

$$W = LL^*\{\cdot\} = \int_{t_0}^{t_1} S(\tau, t_1) S^*(\tau, t_{t_1}) E\{\cdot | F_t\} d\tau$$

and the deterministic controllability matrix $C(t_1, s) : R^n \rightarrow R^n$ is

$$C(t_1, s) = \int_s^{t_1} S(\tau, t_1) S^*(\tau, t_1) d\tau, \quad s \in [t_0, t_1]$$

where the symbol * indicates the matrix ajoint

Theorem 12.1 *Assume that the stochastic system is relatively exactly controllable on $[t_0 \; t_1]$. Then, for arbitrary target $x_1 \in L_2(\Omega, F_t, R^n)$ and $\sigma(\cdot) \in L_2^F([t_0, t_1], R^{n \times n})$, the control*

$$u^0(t) = S^*(t, t_1) E \left[W^{-1}(x_1 - \Phi(t_1, t_0) x_0 - \int_{t_0}^{t_1} \Phi(t_1, s) \sigma(s) d\omega(s) \right.$$
$$\left. - \int_{-h}^{0} d_H \left(\int_{t_0 + s}^{t_0} \Phi(t_1, \tau - s) H(\tau - s, s) u_{t_0} d\tau \right) \right) | F_t \right] \quad (12.2)$$

transfers the system from $x_0 \in R^n$ to $x_1 \in R^n$ at time t_1.

Moreover, among all the admissible controls $u(t)$ transferring the initial state x_0 to the final state x_1 at time $t_1 > 0$, the control $u^0(t)$ minimizes the integral performance index

$$J(u) = E \int_{t_0}^{t_1} \|u(t)\|^2 dt$$

It is well known, that in the theory of dynamical systems with delays in control or in the state variables, it is necessary to distinguish between two fundamental concepts of controllability, namely: relative controllability and absolute controllability (see e.g. [8, 9, 11] for more details).

In this Chapter we shall concentrate on the weaker concept relative controllability on a given time interval $[0, T]$.

12.2 System Description

On the other hand, since for the stochastic dynamical system (1) the state space $L_2(\Omega, F_t, R^n)$ is in fact infinite-dimensional space, we distinguish exact or strong controllability and approximate or weak controllability.

Using the notations given above for the stochastic dynamical system (12.1) we define the following stochastic relative exact and approximate controllability concepts.

Definition 12.1 *The pair $z(t) = \{x(t), u_t\} \in R^n \times L_2^F([t_0, t_1], R^m)$ is said to be the complete state of the system at time t.*

Definition 12.2 *The stochastic system (12.1) is said to be relatively controllable on $[t_0, t_1]$ if, for every complete state $z(t_0)$ and every vector $x_1 \in R^n$, there exists a control $u \in U_{ad}$ such that the corresponding trajectory of the stochastic system satisfies the condition $x(t_1) = x_1$.*

Definition 12.3 *The stochastic dynamical system (12.1) is said to be stochastically relatively exactly controllable on $[t_0, t_1]$ if $R_T(U_{ad}) = L_2(\Omega, F_T, R^n)$ that is, if all the points in $L_2(\Omega, F_T, R^n)$ can be exactly reached from arbitrary initial condition $x_0 \in L_2(\Omega, F_T, R^n)$ at time t_1.*

Definition 12.4 *The stochastic dynamical system (12.1) is said to be stochastically relatively approximately controllable on $[t_0, t_1]$ if $\overline{R_T(U_{ad})} = L_2(\Omega, F_T, R^n)$ that is, if all the points in $L_2(\Omega, F_T, R^n)$ can be approximately reached from arbitrary initial condition $x_0 \in L_2(\Omega, F_T, R^n)$ at time t_1.*

Remark 12.1 From the Definitions 12.1 and 12.2 directly follows, that stochastic relative exact controllability is generally a stronger concept than stochastic relative approximate controllability. However, there are many cases when these two concepts coincide.

Remark 12.2 Since the stochastic dynamical system (12.1) is linear, then without loss of generality in the above two definitions it is enough to take zero initial condition $x_0 = 0 \in L_2(\Omega, F_T, R^n)$.

Remark 12.3 It should be pointed out, that in the case of delayed controls the above controllability concepts essentially depend on the length of the time interval $[t_0, t_1]$.

Remark 12.4 From the form of the controllability operator C_T immediately follows, that this operator is self adjoint.

In the sequel we study the relationship between the controllability concepts for the stochastic dynamical system (12.1) and controllability of the associated deterministic dynamical system of the form

$$dy(t) = \left[A(t)y(t) + \int_{-h}^{0} v(t+s) dH_s(t,s)\right] dt \quad t \in [t_0, t_1] \quad (12.3)$$

$$y(t_0) = y_0$$

where the admissible controls $v \in L_2([t_0, t_1], R^m)$.

Therefore, let us recall the well known lemma concerning relative controllability of deterministic system (12.3).

Lemma 12.1 Klamka [31] *the following conditions are equivalent*:

(i) deterministic system (12.3) is relatively controllable on $[t_0, t_1]$,
(ii) controllability matrix G_T is nonsingular,

Now, for completeness of considerations let us recall well known lemma taken directly from the theory of stochastic processes, which will be used in the sequel in the proofs of the main results.

Lemma 12.2 Mahmudov and Denker [61] and Mahmudovand Zorlu [62] *for every* $z \in L_2(\Omega, F_T, R^n)$, *there exists a process* $q \in L_2^F([0,T], R^{n \times n})$ *such that*

$$C_T z = G_T E z + \int_0^T G_T(s) q(s) dw(s)$$

Taking into account the above notations, definitions and lemmas in the next section we shall formulate and prove conditions for stochastic relative exact and stochastic relative approximate controllability for stochastic dynamical system (12.1).

12.3 Stochastic Relative Controllability

In this section, using lemmas given in Sect. 12.2 we shall formulate and prove the main result of the paper, which says that stochastic relative exact and in consequence also approximate controllability of stochastic system (12.1) is in fact equivalent to relative controllability of associated linear deterministic system (12.3).

Theorem 12.1 *The following conditions are equivalent:*

(i) Deterministic system (12.3) is relatively controllable on $[0, T]$,
(ii) Stochastic system (12.1) is stochastically relatively exactly controllable on $[0, T]$
(iii) Stochastic system (12.1) is stochastically relatively approximately controllable on $[0, T]$.

Proof (i) $\Rightarrow$ (ii) Let us assume that the deterministic system (12.3) is relatively controllable on $[t_0, t_1]$. Then, it is well known (see e.g. [8, 9, 12]) that the symmetric relative deterministic controllability matrix $G_T(s)$ is invertible and strictly positive definite at least for all $s \in [t_0, t_1]$.

12.3 Stochastic Relative Controllability

Hence, for some $\gamma > 0$ we have

$$\langle G_T(s)x, x \rangle \geq \gamma \|x\|^2$$

for all $s \in [t_0, t_1]$ and for all $x \in R^n$. In order to prove stochastic relative exact controllability on $[0, T]$ for the stochastic system (12.1) we use the relationship between controllability operator C_T and controllability matrix G_T given in Lemma 12.2 to express $E\langle C_T z, z \rangle$ in terms of $\langle G_T Ez, Ez \rangle$. First of all we obtain

$$E\langle C_T z, z \rangle = E\left\langle G_T Ez + \int_0^T G_T(s)q(s)dw(s), Ez + \int_0^T q(s)dw(s) \right\rangle$$

$$= \langle G_T Ez, Ez \rangle + E\int_0^T \langle G_T(s)q(s), q(s) \rangle ds \geq \gamma\left(\|Ez\|^2 + E\int_0^T \|q(s)\|^2 ds\right)$$

$$= \gamma E\|z\|^2$$

Hence, in the operator sense we have the following inequality $C_T \geq \gamma I$, which means that the operator C_T is strictly positive definite and thus, that the inverse linear operator C_T^{-1} is bounded. Therefore, stochastic relative exact controllability on $[0, T]$ of the stochastic dynamical system (12.1) directly follows from the results given in [31].

(ii) $\Rightarrow$ (iii) This implication is obvious (see e.g. [31, 38, 39, 45, 47]).

(iii) $\Rightarrow$ (i) Assume that the stochastic dynamical system (12.1) is stochastically relatively approximately controllable on $[t_0, t_1]$, and hence its linear self adjoint controllability operator is positive definite, i.e. $C_T > 0$ [31].

Then, using the resolvent operator $R(\lambda, C_T)$ and following directly the functional analysis method given in [61, 62] for stochastic dynamical systems without delays, we obtain that deterministic system (12.3) is approximately relatively controllable on $[t_0, t_1]$.

However, taking into account that the state space for deterministic dynamical system (12.3) is finite dimensional, i.e. exact and approximate controllability coincide [31], we conclude that deterministic dynamical system (12.3) is relatively controllable on $[t_0, t_1]$.

Remark 12.7 Finally, it should be pointed out, that using general method given in the monograph [31], for stochastically relatively approximately controllable dynamical systems it is possible to formulate the admissible controls $u(t)$, defined for $t \in [t_0, t_1]$ and transferring given initial state x_0 to the desired final state x_T at time t_1.

In this Chapter sufficient conditions for stochastic relative exact controllability for linear stationary finite-dimensional stochastic control systems with multiple constant point delays in the control have been formulated and proved. These conditions extend to the case of one constant point delay in control, known stochastic exact controllability conditions for dynamical control systems without delays recently published in the papers [8, 38, 39, 45, 46, 47].

Finally, it should be pointed out, that using the standard techniques presented in the monograph [31] it is possible to extend the results presented in this paper for more general nonstationary linear stochastic control systems with many time variable point delays in the control. Moreover, the extension for stochastic absolute exact controllability and stochastic absolute approximate controllability in a given time interval is also possible.

References

1. Babiarz, A., Bieda, R., Jaskot, K., Klamka, J.: The Dynamics of the human arm with an observer for the capture of body motion parameters. Bull. Pol. Acad. Sci. Tech. Sci. **61**(4), 955–971 (2013)
2. Babiarz, A., Czornik, A., Klamka, J., Niezabitowski, M.: The selected problems of controllability of discrete-time switched linear systems with constrained switching rule. Bull. Pol. Acad. Sci. Tech. Sci. **63**(3), 657–666 (2015)
3. Balachandran, K., Dauer, J.P.: Controllability of nonlinear systems via fixed point theorems. J. Optim. Theory Appl. **53**, 345–352 (1987)
4. Balachandran, K., Dauer, J.P.: Controllability of nonlinear systems in Banach spaces. Survey J. Optim. Theory Appl. **115**, 7–28 (2002)
5. Balachandran, K., Park, D.G., Manimegalai, P.: Controllability of second order integro differential evolution systems in Banach spaces. Comput. Mathe. Appl. **49**, 1623–1642 (2005)
6. Breuer, H.P., Petruccione, F.: The Theory Of Open Quantum Systems. Oxford University Press, USA (2002)
7. Dong, D., Petersen, I.R.: Quantum control theory and applications: a survey. IET Control Theory Appl. **4**(12), 2651–2671 (2010)
8. Ficak, B., Klamka, J.: Stability criteria for a class of stochastic distributed delay systems. Bull. Pol. Aca. Sci. Tech. Sci. **61**(1), 221–229 (2013)
9. Fu, X.: Controllability of abstract neutral functional differential systems with unbounded delay. Appl Mathe. Comput. **141**, 299–314 (2004)
10. Gawron, P., Klamka, J., Miszczak, J., Winiarczyk, R.: Extending scientific computing system with structural quantum programming capabilities. Bull. Pol. Acad. Sci. Tech. Sci. **58**(1), 77–88 (2010)
11. Ge, S.S., Sun, Z., Lee, T.H.: Reachability and controllability of switched linear discrete-time system. IEEE Trans. Autom. Control AC **46**(9), 1437–1441 (2001)
12. Hernandez, E., Henriquez, H.R.: Impulsive partial neutral differential equations. Appl. Mathe. Lett. **19**, 215–222 (2006)
13. Hernandez, E., Henriquez, H.R., Rabello, M.: Existence of solutions for impulsive partial neutral functional differential equations. J. Mathe. Analysis Appl. **331**, 1135–1158 (2007)
14. Kaczorek, T.: Positive 1D and 2D Systems. Springer-Verlag, London (2002)
15. Kaczorek, T.: Rechability and controllability to zero of cone fractional linear systems. Archives Control Sci. **17**(3), 357–367 (2007)
16. Kaczorek, T.: Fractional positive continuous-time linear systems and their reachability. Int. J. Appl. Mathe. Comput. Sci. **18**(2), 223–228 (2008)

17. Kaczorek, T., Rogowski K.: Fractional linear systems and electrical circuits, studies in systems. In: Decision and Control, vol. 13. Springer (2015)
18. Kaczorek, T.: Positive fractional 2D continuous discrete time linear systems. Bull. Pol. Acad. Sci. Tech. Sci. **59**(4), 575–579 (2011)
19. Kaczorek, T.: Positive switched 2D linear systems described by the Roesser models. Euro. J. Control **18**(3), 239–246 (2012)
20. Kilbas, A.A., Srivastava, H.M., Trujillo, J.J.: Theory and Applications of Fractional Differential Equations. Elsevier, Amsterdam (2006)
21. Klamka, J.: Relative controllability and minimum energy control of linear systems with distributed delays in control. IEEE Trans. Autom. Control AC **21**(4), 594–595 (1976)
22. Klamka, J.: Controllability of linear systems with time-variable delays in control. Int. J. Control **24**(6), 869–878 (1976)
23. Klamka, J.: On the controllability of linear systems with delays in the control. Int. J. Control **25**(6), 875–883 (1977)
24. Klamka, : Absolute controllability of linear systems with time-variable delays in control. Int. J. Control **26**(1), 57–63 (1977)
25. Klamka, J.: Relative and absolute controllability of discrete systems with delays in control. Int. J. Control **26**(1), 65–74 (1977)
26. Klamka, J.: Minimum energy control of discrete systems with delays in control. Int. J. Control **26**(5), 737–744 (1977)
27. Klamka, J.: Relative controllability of nonlinear systems with distributed delays in control. Int. J. Control **28**(2), 307–312 (1978)
28. Klamka, J.: Relative controllability of infinite-dimensional systems with delays in control. Syst. Sci. **4**(1), 43–52 (1978)
29. Klamka, J.: Minimum energy control of 2D systems in Hilbert spaces. Syst. Sci. **9**(1–2), 33–42 (1983)
30. Klamka, J.: Sterowalność Układów Dynamicznych. PWN, Warszawa (1990). (in Polish)
31. Klamka, J.: Controllability of Dynamical Systems. Kluwer Academic, Dordrecht (1991)
32. Klamka, J.: Controllability of dynamical systems—a survey. Arch. Control Sci. **2**(3–4), 281–307 (1993)
33. Klamka, J.: Constrained controllability of nonlinear systems. J. Mathe. Analysis Appl. **201**(2), 365–374 (1996)
34. Klamka, J.: Schauder's fixed point theorem in nonlinear controllability problems. Control Cybern. **29**(3), 377–393 (2000)
35. Klamka, J.: Constrained exact controllability of semilinear systems. Syst. Control Lett. **47**(2), 139–147 (2002)
36. Klamka, J.: Positive controllability of positive dynamical systems. Proc. 20th Ann. Am. Control Conf. **1–6**, 4632–4637 (2002). Anchorage
37. Klamka, J.: Constrained controllability of semilinear systems with multiple delays in control. Bull. Pol. Acad. Sci. Tech. Sci. **52**(1), 25–30 (2004)
38. Klamka, J.: Approximate constrained controllability of mechanical systems. J. Theor. Appl. Mech. **43**(3), 539–554 (2005)
39. Klamka, J.: Stochastic controllability of linear systems with delay in control. Bull. Pol. Acad. Sci. Tech. Sci. **55**(1), 23–29 (2007)
40. Klamka, J.: Stochastic controllability of linear systems with state delays. Int. J. App. Mathe. Comput. Sci. **17**(1), 5–13 (2007)
41. Klamka, J.: Stochastic controllability of systems with variable delay in control. Bull. Pol. Acad. Sci. Tech. Sci. **56**(3), 279–284 (2008)
42. Klamka, J.: Constrained controllability of semilinear systems with delayed controls. Bull. Pol. Acad. Sci. Tech. Sci. **56**(4), 333–337 (2008)
43. Klamka, J.: Stochastic controllability and minimum energy control of systems with multiple delays in control. Appl. Mathe. Comput. **206**(2), 704–715 (2008)

References

44. Klamka, J.: Controllability of fractional discrete-time systems with delay. Zeszyty Naukowe Politechniki Śląskiej seria Automatyka **151**, 67–72 (2008)
45. Klamka, J.: Stochastic controllability of systems with variable delay in control. Bull. Pol. Acad. Sci. Tech. Sci. **56**(3), 279–284 (2008)
46. Klamka, J.: Constrained controllability of semilinear systems with delayed controls. Bull. Pol. Acad. Sci. Tech. Sci. **56**(4), 333–337 (2008)
47. Klamka, J.: Stochastic controllability and minimum energy control of systems with multiple delays in control. Appl. Mathe. Comput. **206**(2), 704–715 (2008)
48. Klamka, J.: Constrained controllability of semilinear systems with delays. Nonlinear Dyn. **56**(1–2), 169–177 (2009)
49. Klamka, J.: Stochastic controllability of systems with multiple delays in control. Int. J. Appl. Mathe. Comput. Sci. **19**(1), 39–47 (2009)
50. Klamka, J.: Controllability and minimum energy control problem of fractional discrete-time systems, chapter in monograph. In: Baleanu, D., Guvenc, Z.B., Tenreiro Machado J.A. (eds.) New Trends in Nanotechnology and Fractional Calculus, pp. 503–509. Springer-Verlag, New York (2010)
51. Klamka, J.: Controllability of dynamical systems. A Survey. Bull. Pol. Acad. Sci. Tech. Sci **61**(2), 221–229 (2013)
52. Klamka, J., Czornik, A., Niezabitowski, M.: Stability and controllability of switched systems. Bull. Pol. Acad. Sci. Tech. Sci. **61**(3), 547–554 (2013)
53. Klamka, J.: Constrained controllability of second order dynamical systems with delay. Control Cybern. **42**(1), 111–121 (2013)
54. Klamka, J., Babiarz, A., Niezabitowski, M.: Banach fixed-point theorem in semilinear controllability problems—a survey. Bull. Pol. Acad. Sci. Tech. Sci. **64**(1), 21–23 (2016)
55. Klamka, J., Socha, L.: Some remarks about stochastic controllability, IEEE Transactions on Automatic Control AC-22. **5**, 880–881 (1977)
56. Klamka, J., Socha, L.: Some remarks about stochastic controllability for delayed linear systems. Int. J. Control. **32**(3), 561–566 (1980)
57. Klamka, J., Wyrwał, J.: Controllability of second-order infinite-dimensional systems. Syst. Control Lett. **57**(5), 386–391 (2008)
58. Klamka, J., Sikora, B.: On constrained stochastic controllability of dynamical systems with multiple delays in control. Bull. Pol. Acad. Sci. Tech. Sci. **60**(2), 301–306 (2012)
59. Lin, Y., Tanaka, N.: Nonlinear abstract wave equations with strong damping. J. Mathe. Anal. Appl. **225**(1), 46–61 (1998)
60. Liu, X., Wilms, A.R.: Impulsive controllability of linear dynamical systems with applications to maneuvers of spacecraft. Mathe. Problems Eng. **2**, 277–299 (1996)
61. Mahmudov, N.I., Denker, A.: On controllability of linear stochastic systems. Int. J. Control **73**(2), 144–151 (2000)
62. Mahmudov, N.I., Zorlu, S.: Controllability of nonlinear stochastic systems. Int. J. Control **76**(2), 95–104 (2003)
63. Naito, K.: Controllability of semilinear control systems dominated by the linear part. SIAM J. Control Optimization **25**(3), 715–722 (1987)
64. Ostalczyk, P.: The non-integer difference of the discrete-time function and its application to the control system synthesis. Int. J. Sys. Sci. **31**(12), 1551–1561 (2000)
65. Park, J.Y., Balachandran, K., Arthi, G.: Controllability of impulsive neutral integrodifferential systems with infinite delay in Banach spaces. Nonlinear Anal. Hybrid Syst. **3**, 184–194 (2009)
66. Peichl, G., Schappacher, W.: Constrained controllability in Banach spaces. SIAM J. Control Optimization **24**(6), 1261–1275 (1986)
67. Qiao, Y., Cheng, D.: On partitioned controllability of switched linear systems. Automatica **45**(1), 225–229 (2009)
68. Robinson, S.M.: Stability theory for systems of inequalities. Part II. Differentiable nonlinear systems. SIAM J. Numer. Anal. **13**(4), 497–513 (1976)

69. Seidman, T.I.: Invariance of the reachable set under nonlinear perturbations. SIAM J. Control Optimization **25**(5), 1173–1191 (1987)
70. Sakthivel, R., Mahmudov, N.I., Kim, H.J.: On controllability of second-order nonlinear impulsive differential systems. Nonlinear Anal. **71**, 45–52 (2009)
71. Sikora, B., Klamka, J.: New controllability criteria for fractional systems with varying delays. In: Babiarz, A., Czornik, A., Klamka, J., Niezabitowski, M. (eds.) Theory and Applications of Non-integer Order Systems. Lecture Notes in Electrical Engineering, vol. 407. Springer Verlag (2017)
72. Son, N.K.: A unified approach to constrained approximate controllability for the heat equations and the retarded equations. J. Mathe. Anal. Appl. **150**(1), 1–19 (1990)
73. Sun, Z., Ge, S.S., Lee, T.H.: Controllability and reachability criteria for switched linear system. Automatica **38**(5), 775–786 (2002)
74. Sun, Z., Ge, S.S.: Switched Linear Systems—Control and Design. Springer, New York (2004)
75. Sun, Z., Zheng, D.Z.: On reachability and stabilization of switched linear control systems. IEEE Trans. Autom. Control AC **46**(2), 291–295 (2001)
76. Węgrzyn, S., Klamka, J., Bugajski, S.: Foundation of quantum computing. Part 1. Archiwum Informatyki Teoretycznej i Stosowanej **13**(2), 97–142 (2001)
77. Węgrzyn, S., Klamka, J., Bugajski, S.: Foundation of quantum computing. Part 2, Archiwum Informatyki Teoretycznej i Stosowanej **14**(2), 93–106 (2002)
78. Węgrzyn, S., Klamka, J., Znamirowski, L., Winiarczyk, R., Nowak, S.: Nano and quantum systems of informatics. Bull. Pol. Acad. Sci. Tech. Sci. **52**(1), 1–10 (2004)
79. Xie, G., Wang, L.: Controllability and stabilizability of switched linear systems. Syst. Control Lett. **48**(2), 135–155 (2003)
80. Zhou, H.X.: Controllability properties of linear and semilinear abstract control systems. SIAM J. Control. Optimization **22**(3), 405–422 (1984)

Printed by Printforce, the Netherlands